RECONSIDERING IAN MCHARG: THE FUTURE OF URBAN ECOLOGY

IGNACIO F. BUNSTER-OSSA

American Planning Association
Planners Press

Making Great Communities Happen

Chicago | Washington, D.C.

Copyright © 2014 by the American Planning Association
205 N. Michigan Ave., Suite 1200, Chicago, IL 60601-5927
1030 15th St., NW, Suite 750 West, Washington, DC 20005-1503
www.planning.org
ISBN 978-1-61190-123-8
Library of Congress Control Number 2013953171
Printed in the United States of America
All rights reserved

Contents

Foreword	vii
Introduction	xi
Reading This Book	xvii
Acknowledgments	xxi
Dedication	xxiii

Chapter 1. *Design with Nature*: Promises and Pitfalls	**1**
Ecology as a Promised Land	4
The Layer Cake	5
The Plan for the Valleys	7
Abetting Sprawl, Failing the Inner City	10
Design with *Which* Nature?	13

Chapter 2. The American Wilderness: An Evanescing Myth	**19**
Wild Lands That Hardly Were	21
Infrastructure Rising	24
Wilderness Subsumed	27

Chapter 3. Cities: Our Abode	**31**
Densifying the American Dream	34
Up, Up, and (Not) Away	39

Chapter 4. Building, Dwelling, Greening   45
Breathing Green   46
From Design with Nature to the Granite Garden   49
GreenPlan Philadelphia   53

Chapter 5. From Green to White: Ecology as a Design Ethic   67
Body and Soul of the Baroque   68
Fargo 365   79
Leibniz and Compossibility   83
A "Unity of the Arts"   86

Chapter 6. Localism: A Participatory Ecology   91
Localism as an "Eco-Democratic" Circumstance   93
The Seattle Question   94
Of Wharves and Gardens   101

Chapter 7. On Public Art   109
Site Specificity   112
Site Specific Utility   115
Site Specific *Ecological* Utility   121

Chapter 8. Dallas: In Search of an Urban Future   129
Planning Background   130
In Horse Country, A Magic Hoof   132
Making It Happen   136
Public Art   141
Economic Development   143

Chapter 9. Toward a Climax City   147
City as Landscape   149
Community as Park   155
Building as Garden   160

Chapter 10. Beyond, Ahead	**167**
Delivering on Densification	168
Return on Investment	172
A Role for Government	174
A Final Word on Ian L. McHarg	179
End Notes	**181**
Bibliography	**187**
Photo Credits	**192**
Index	**193**

Foreword

By Harriet Tregoning

Cofounder, Smart Growth Network and
former Secretary of Planning
for the State of Maryland

In many ways, we are at a critical turning point in civilization. It is not just the moment when the majority of human beings began to make their homes in cities. It is also the moment when the rapidly evolving knowledge economy has ushered us more or less out of the Industrial Age. Yet, many of our concepts about what constitutes the *good life* in the western world and—through western media—much of the rest of the world, were being formed in the debates about democracy and citizenship that took place much longer ago, in the early years of a fledgling American democracy, in the cusp between an agrarian way of life that had lasted for millennia and a dawning Industrial Age.

The two competing views on how to shape the young American democracy are represented by Thomas Jefferson, a gentleman farmer, plantation owner, and slaveholder, who disliked the professional politician—the man without his own land to live by, and Alexander Hamilton, a self-made man who married into wealth; loved efficiency, order, and organization; and favored a strong central government acting in the interests of commerce and industry. Jefferson advocated a decentralized agrarian republic. He appreciated the value of a strong central government only in foreign relations. Hamilton's objective was more efficient organization, while Jefferson once said, "I am not a friend to a very energetic government."[1]

For Jefferson, a society based on self-sufficiency and a deep, almost spiritual, understanding of the rhythms and processes of nature begat justice. He espoused a republicanism founded on a political and personal independence. For the most part, Jefferson disdained cities. "The mobs of great cities add just so much to the support of pure government, as sores do to the strength of the human body."[2]

Hamilton saw civic investment as an essential ingredient in the development of the citizen. The foundation of a just society was not based on agrarian values but rather on the creation and endurance of strong social and political institutions which could guide moral development; planning and organization was inherently urban.

More than 200 years later, the contrast between the urban and the rural experience and how they affect theories of democracy continue to play out in contemporary politics and policy. The relative virtues of "urban versus rural" color our perceptions of what constitutes the *good life* and whether terms like "suburban" and "sprawl" are more pejorative or aspirational. With the prediction that by mid-21st century, three-quarters of the human race will live in cities, we have to consider what aspects of the urban and rural experience we are to incorporate into our collective and community identities, aspirations, and physical spaces. In *Reconsidering Ian McHarg: The Future of Urban Ecology*, Ignacio F. Bunster-Ossa has offered to be our guide.

Initially, suburbia held out the tantalizing promise of combining the virtues of both urban and rural life. In his seminal work, *Design with Nature*, McHarg inveighed against unthinking development and despoliation and in favor of protecting the natural environment, primarily by gaining an understanding of ecology as a basis for planning and design. In theory and in practice, McHarg offered a way to harmoniously develop outside cities, in concert with the natural ecology. His vision of the good life was firmly in the agrarian tradition, and his practice and teachings helped to popularize an entire category of *greener* suburban development such as Prairie Crossing far outside Chicago.

Indeed, after World War II, suburban life generally promised a compelling reconciliation—the best of both urban and rural: closer to nature, with the green lawn faintly recalling the English manor house; reasonably priced housing on cheap land far from the city center; privacy, not the overcrowding of the tenement; the safe, low density of population enabling instant distinction of neighbor versus stranger; family-friendly subdivisions free from speeding traffic; easy car commutes into downtowns or other job centers; healthful lifestyles with fresh air, and big back yards.

Foreword

Over time, however, the true countryside moved farther away and the faux countryside—the suburban countryside—developed. As mounting numbers opted to live in the suburbs, they often experienced diminishing returns. And, as suburbia spread farther from cities, easy commutes inexorably lengthened and became unpredictable; instead of better health, suburban life meant hours each day behind the wheel of a car, leaving little time for exercise and less time for family or getting to know neighbors; low density neighborhoods and disconnected street networks left few daily destinations within walking distance; residential affordability was eroded by the need for every adult to have a car; and a big yard often required a high level of maintenance and water, while fertilizer and pesticide further undermined claims to a more ecologically benign impact.

Can some of those formative Jeffersonian ideals now be transferred to the Hamiltonian sphere of cities? Can environmental stewardship, ecological consciousness, and both knowledge of and respect for nature be fostered in an ever urbanizing society? What about democracy and the ability of people to organize and act as citizens with concern for the welfare of each other and the larger society?

Professor Michael Sandel, a political philosopher at Harvard University, addressed this issue at a national planning meeting in 2011.[3] A growing threat to public life, according to Sandel, is the increasing atomization of our society. As different demographic and socioeconomic groups isolate themselves in their own spaces—their own neighborhoods, restaurants, and clubs—they withdraw from public places. Why should we be concerned about this? "Because there is a deep civic loss," Sandel warned. "And public institutions (such as libraries, museums, and parks) cease to be places where people from different walks of life encounter each other." There are no longer any opportunities for "informal schools of citizenship and virtue." Sandel recognized that with changing preferences for city living and the re-emergence of cities around the globe, design professionals must be challenged to create cities that are hospitable to civic virtue and democratic citizenship.

In *Reconsidering Ian McHarg*, Bunster-Ossa brings together these many threads, giving us his perspective on the compelling ecological virtues of urbanity and the compelling human need to be proximate to a nature-heeled urban landscape. Bunster-Ossa recognizes McHarg's prescience in employing natural systems to provide valuable ecological services, but advances their application to densifying cities, reinforcing their multi-tasking, community-building utility. Bunster-Ossa emphasizes the critical value of culture and community life that brings people of disparate backgrounds, incomes, ethnicities, occupations, and viewpoints together, reminding us that the words *citizen* and *city* have a common root in the term *civitas*. Most importantly, Bunster-Ossa underscores that what binds us to a place are the private and shared moments of recognition, connection, and sublimity that come from a built environment that is healthy, legible, and artful.

Many nations, including the United States, are experiencing a profound shift in demographics, with longer-lived seniors, young people who are forming families later or not at all, fewer households with children, and many more single-person households at every age. Moreover, a generational shift seems to be underway in western nations, with a stronger preference for urban living, along with a broad global movement to cities for those in search of greater economic opportunity and mobility.

With this volume, Bunster-Ossa addresses the urgent need of cities everywhere not only to create humane and welcoming places for the majority of humanity, but also to give us greater capacity to come together as neighbors and as global citizens to tackle society's greatest problems and to truly design with nature, especially with our increasingly urban nature.

Introduction

Addressing a national gathering of landscape architects in Portland in 1998, former Senator Gaylord Nelson, founder of Earth Day, underscored the imperative of steering humanity toward a sustainable future. He exhorted the group to think globally, especially with respect to the preservation of open space, agricultural lands, and natural resources. He posed the challenge of having to provide for the next doubling of the world population—food, shelter, clean water, clean air, energy, and so on—emphasizing how, in meeting this call, landscape architects had a unique role to play.

This was sweet music to the attendees. Landscape architects, after all, are trained in matters of land stewardship and resource conservation. But my mind drifted mid-speech. I began to transpose the Senator's call to the urban environment, pondering what cities might look like after a doubling of the world's population, the U.S. population in particular. Would 600 million Americans require the same proportion of urban land as the first 300 hundred million? Double the areas of Los Angeles, Phoenix, Dallas, Atlanta? Not likely, of course; that would defy any measure of sustainability. Any plausible answer made placing more people per unit of urban land—*densification*—inevitable.

It is hardly an exaggeration to suggest that, barring global warfare, deadly pandemics, or the strike of a wayward meteor, the quality of human life on earth will, to a large extent, depend on the quality of our cities: on their geographic reach, the energy they consume, and the environment they engender. For city planners, urban designers, architects, and landscape architects, this means not merely mitigating the impacts of human settlement upon nature, but also creating a new *urban nature*, one that improves every global and local measure of environmental quality, community well-being, and personal health.

Significant effort has been expended over the past decades in attempts to mitigate the environmental impact of urban sprawl. We must now build on this foundation and train our sights upon core urban areas: regional centers, downtowns, mixed-use districts, transit-oriented developments, redeveloped malls, and transportation hubs. We must do this not merely to create cities that are more sustainable, but rather to turn them—over suburban alternatives—into marketable and denser *environments of choice*.

In examining the issue of environmental quality in the context of urban densification, three fundamental necessities arise:

- *Green Infrastructure*, as the means to integrate a "working nature" in the urban midst;
- *Localism*, as the means to reaffirm the value of culture and community life; and
- *Public Art*, as the means to exact from everyday life a measure of rooted meaning, beauty, and sublimity.

These are the strands that underpin the planning and design arguments that follow. While the subjects of green infrastructure, localism, and public art have been well discussed individually in the planning and design fields, and case studies abound in each instance, comparatively little has been advanced in terms of folding them together holistically as an urban ecology. This book fills that void by presenting ecology as a unifying and binding agent—as a science, to be sure, but more importantly as an *ethic* that enfolds matter and spirit as planning and design concerns.

Owing to the importance of ecology as an underlying concern, this book forcibly engages the legacy of Ian L. McHarg. As a landscape architect, educator, and practitioner, McHarg introduced the integration of ecology into the city planning and design discourse, changing the method by which urban development was molded within the matrix of nature. His legacy is as vital to landscape architecture as that of Frederick Law Olmsted who, with a handful of peers, established the profession in 1900.

INTRODUCTION

Olmsted's outsize legacy derives in part from the planning and design of major urban parks such as New York City's Central and Prospect Parks. In creating such public landscapes, Olmsted aimed to shape the way cities could grow while improving public health through access to recreation and places for social exchange. He also pioneered the use of natural systems as cleansing agents in urban environments, most notably in the "fens" of Boston's Back Bay, a work that ushered the concept of "ecological service" a century before the concept gained global traction as a way to accord nature infrastructural value.

McHarg's legacy was hewn decades later through an impassioned plea to protect the natural environment against despoliation and abuse, principally through an understanding of ecological process as a basis for planning and design. In his seminal publication, *Design with Nature*, he lays out both thesis and method. The book was published in 1969 as a summary of "Man and Environment," the emblematic course on ecology and urban development he established at the University of Pennsylvania's Department of Landscape Architecture and Regional Planning upon becoming chair in 1959. His words advocate ways for regions to grow rationally, defensibly and deterministically in the interest of health—human and wild. They also forewarn the consequences of building ignorantly in ecologically sensitive environments. (The first chapter of *Design with Nature* explains the ecology of the New Jersey barrier islands. It points to the inherent hazard of building towns upon the dune environment, a matter that acquired renewed if tragic relevance in 2012 after Hurricane Sandy laid waste to large sections of the New Jersey coastline.)

For decades, *Design with Nature* was mandatory reading for anyone entering a landscape or planning program, and it remains a milestone in the professional literature. Current ideas about sustainability and permaculture can be directly traced to McHarg's work and pedagogy, as can the practice of assessing environmental impact, as mandated by the National Environmental Protection Act (NEPA).

Unfortunately, McHarg's ecological planning method was, and to a large extent continues to be, applied to the development of greenfield sites. In providing a method for developing the hinterland without unduly impacting valuable ecological resources, *Design with Nature*, in effect, facilitated the spread of suburbia across the American landscape. Here was a method that could deposit brand-new homes, roads and businesses in nature's midst under the well-intentioned mantle of environmentalism. In this sense, *Design with Nature* is a misnomer: it predicated, rather, design *within* nature.

Generations of planners and landscape architects have in this way abetted greenfield sprawl and created, through consequent vehicular use and emissions, a far greater environmental problem than anything *Design with Nature* ever attempted to solve. This outcome could not have been predicted 50 years ago. In retrospect, however, the confluence of government policies, market forces, and social trends leading to it are all too clear as abetting factors. Fifty years ago environmentalism was on the rise. The federal government was funding highways full-throttle, opening vast areas of rural land for development. People wanted their own slice of nature via a single-family house and yard, ideally near a wooded lot or stream, and developers and land owners were eager to oblige.

The poster child of this phenomenon is The Woodlands New Community, a wooded site abutting Interstate 45 approximately 30 miles north of Houston, Texas. Planned in the early 1970s and supported by federal land use policies, the site was beset by hydrological constraints, but McHarg and his team devised a plan that leveraged them into a profitable venture. The preservation of buffered streams, constructed swales, retention ponds, and aquifer recharge areas allowed the new development to maintain the extant water balance, avoiding expensive engineered solutions and sidestepping potential permitting difficulties. Single-family homes were clustered and their footprints and yards minimized in an effort to preserve the tree canopy—a mix of pine and hardwood forest. The development drew throngs of homebuyers intent on living amid green placidity, proof positive of the virtues of the ecological planning method.

And yet despite a sound ecological foundation, the development is hardly sustainable. Most residents must drive for services and either take to the wheel or suffer long bus commutes to get to work in downtown Houston. The Metropolitan Transportation Authority of Harris County is set to build a six-mile northbound extension of the regional light rail in the direction of The Woodlands, but its last stop will fall 20 miles short, keeping a majority of the community's approximately 60,000 residents road-

INTRODUCTION

way-bound for decades to come. One has to wonder: what kind of a place would Houston be today if even half of those who flocked to the wooded hinterland had remained in the city? What carbon emission savings and public health benefits would have accrued by now?

A reflection on McHarg's work is offered as a way to highlight the significant differences between the practice of ecological planning and design then and now. I do so as McHarg's successor at WMRT, the firm he cofounded in 1963 with David Wallace (a city planner), William Roberts (a landscape architect), and Tom Todd (an architect). (In the early 1980s a rift occurred among the four principals over Pardisan, a large environmental park in Iran, funded by the Shah and developed under McHarg's insistent guidance. Following the Iranian revolution, the Ayatollahs stopped the project; they also withheld payment for the work, a matter that was ultimately resolved by the International Court of Justice at The Hague. McHarg parted ways with his partners as a result of the debacle, becoming an independent consultant while remaining as chair of the Department of Landscape Architecture and Regional Planning at the University of Pennsylvania. With the "M" gone, WRT emerged as the succeeding entity.)

My life and work have been shaped by McHarg's work and influence in profound ways. I have done work that has been the very embodiment of McHarg's template, planning and designing projects where development was fitted ever so carefully into the rural fringe, satisfying every environmental law in the book yet enabling the very sprawl that undermines the value of that work. I have thus come to experience first-hand the limitations of McHarg's approach to planning and design, which include matters of social equity and community individuation that were hardly on the radar during his rise to fame.

Over the past decades, WRT, McHarg's successor firm, has evolved in parallel with the shifts in policy and practice related to sustainable urbanism. The evolutionary process has produced waves of introspection regarding the firm's founding ethos: What does *Design with Nature* mean now? What currency does it have in the context of development densification? What does ecological process imply in community-bound spaces? And, what planning and design methods might points to suitable answers? These questions are vital to any planning and design practice. But they are equally vital to society at large.

Reading this Book

Using *Design with Nature* as a point of departure in any discussion on city planning and urban design carries a burden: addressing the concept of nature as McHarg considered it and, as important, how it resides in the American psyche.

Chapter 1 addresses the first of these points, both as a promise and pitfall to "designing with nature." For the uninitiated, the first chapter also lays out the basic tenets of McHarg's ecological method, how it was employed as a land-planning tool, and to what effect.

Chapter 2 addresses the second point: how the idea of nature has been embraced as a cultural construct, namely the myth of a boundless wilderness. The idea of nature is also examined in the context of infrastructure and the many ways it has shaped the landscape, leading, ultimately to the managed compartmentalization of wild lands within the urban domain.

Chapter 3 summarizes the post-myth reality of the urban domain from an environmental standpoint, especially as related to public health. Also discussed are the trends pointing toward development densification and, as a corollary, the value of green infrastructure as an enabling agent of urban health and wellness.

Chapter 4 focuses on the elements of green infrastructure and the ways it can improve the urban environment. The landmark study, Green-Plan Philadelphia, is presented as a guide. Focus is given to carbon sequestration, climate mitigation, stormwater management, watershed restoration, and active mobility. The chapter concludes with a brief examination of Anne Spirn's thesis of urban ecology as the city-focused antipode to McHarg's *Design with Nature*.

Chapter 5 examines ecology from an ethical standpoint and its meaning in the context of urban planning and design. The question is viewed through the philosophy of G.W. Leibniz and its critique by French philosopher Gilles Deleuze, especially as it concerns matter and spirit as a syncretism by which to understand and develop place-specific ecologies. Baroque architecture is examined as an expression of such syncretism, providing also a guide for the pursuit of an ecologically-based "unity of the arts."

Chapter 6 discusses the central role of community identity and engagement in the derivation of place ecologies and, within this context, what the work of the designer entails: Is the open, democratic process of decision-making compatible with the goal of design excellence?

Chapter 7 introduces the necessity of art as a means to elevate green infrastructure and localism as the ore of place-poetics; that is, the means by which urban environments can induce wonder, provide inspiration, and affect rootedness as a derivative of biocultural specificity. The progression of public art from the 1960s onwards is reviewed as a way to underscore the emergence and ultimate potential of the genre as a form of civic infrastructure.

Chapter 8 examines the relevance of the preceding concepts in the context of Dallas, Texas, a city that has made strides toward attaining a sustainable future. A detailed review of the Trinity River Corridor Project (TRCP) is provided as an example of the ongoing effort to recast the core area of the city as a transit- and recreation-oriented, mixed use green and art-inspired environment.

Chapter 9 discusses the ultimate aim of green infrastructure, localism, and public art, namely the ecological equivalent of a climax urban environment, or sustaining urban "end-state." "Climax City" is introduced as a theoretical planning and design model, scaled around three constructs: City as Landscape, Community as Park, and Building as Garden.

To end, **Chapter 10** posits the possible adoption of a "climax cities" program as an environmental, economic, and national security priority. Discussed are issues of densification, supply and demand, the return on investment of green infrastructure as a densification catalyst, and the role of government toward effecting change. The conclusion is clear and simple—the urgency with which McHarg approached the conservation of land from an ecological standpoint must now be turned toward the densification of American cities, with a new view of ecology applied toward a sustainable future.

In conceiving this work, I admit without apology a bias toward landscape architecture. This bias is informed by the notion that "environment" suffuses every aspect of the city-building enterprise, one that requires a holistic mindset such as McHarg professed and taught. It is a bias, too, that regards the disposition and function of public spaces as weighing as much or more on the quality of urban environment as the totality of buildings therein. By extension, I also profess a bias against the traditional dichotomy between "building" and "landscape," favoring instead a more fluid exchange between "inside" and "outside," one in which light, air, water, energy, vegetation, food production, social exchange, and, yes, art exist as mediating systems. These conceptions have *Design with Nature* as a source; yet they reach beyond this most inspired call for rational and informed design by turning the focus upon the one wilderness that has yet to attain ecological health: our cities.

Ian L. McHarg died in 2001. A Quaker service was held at a Meeting Hall near his home set deep in the pastoral setting of Chester County, outside of Philadelphia. In true Quaker fashion, people spoke of his life, achievements, and impact upon individuals and society at large. Many remained silent, introspection stirring, waiting, perhaps, for another opportunity to offer a measured view of his legacy in the context of a rapidly changing world.

Acknowledgments

This book would not have been undertaken without McHarg's legacy as an endowment with which to imagine and work toward a better world. I owe him much as a teacher and mentor, including the obligation to question and critique the world as it is and posit without apology by what methods it might be changed. This book would not have been undertaken had it not also been for Daniel S. Friedman, dean of the University of Washington's College of the Built Environment who, during the interview process for the design of the Seattle Central Waterfront in 2010, posed a timely and provocative question about the presumed incompatibility between community participation and design excellence. I duly thank him as well as members of the project's official interview committee, which deemed my team's proposal worthy of an interview but not worthy enough to merit selection. Had our team received the commission, the impetus to turn our proposal's ethical underpinnings into a book would surely have dissipated.

I acknowledge the contribution of Hideo Sasaki, who encouraged me to apply to the Harvard University Loeb Fellowship and backed it up with a letter of recommendation. I devoted the Fellowship almost entirely to the study of ecological ethics, a subject that deeply informed the answer to Friedman's question. To Clive Dilnot, inspiring ethics teacher and mentor: thank you, wherever you are. I thank Dan Rose, teacher, friend, artist, and kindred spirit who provided, since the early days of this venture, unqualified support and advice. Without his encouragement and insightful critique the writing could well have wavered and possibly altogether stopped.

Several friends and colleagues provided helpful advice and information: Jonathan Barnett on the need to address McHarg's legacy; William H. Roberts and Teresa Moore on the *Plan for the Valleys*; David Witham on Fargo, N.D., and his winning urban design competition; Anne Satterthwaite and Elizabeth McKeown on the history of Georgetown in Washington, D.C., and the community's sense of identity, respectively; Mindy Taylor Ross and Kevin Osburn on the Indianapolis Cultural Trail; Don Raines, Gail Thomas, and Karen Walz on Dallas and the Trinity River Corridor; Jody Pinto and Brad Goldberg on public art; Kevin Burke on the Atlanta Beltline; Anne Whiston Spirn on the West Philadelphia Project; and Howard Neukrug on Philadelphia's Green City, Clean Waters program.

I thank my father for inspiring me to view the world from 10,000 feet (he was a commercial pilot) and my mother for instilling an appreciation for art (she was an art historian). I deeply thank my wife Sylvia Palms. As a landscape architect and graduate of Penn's program in landscape architecture, she provided steadfast support while shouldering a disproportionate share of our child-rearing responsibilities. That she also immersed herself in the book as a hard-nosed critic I will forever cherish.

Finally, I owe gratitude to the editors of APA Planners Press, who saw inherent value in ideas not yet cohered and, with a kind and unerring voice, steered me towards the words that lie ahead.

Dedication

To the Bees in my bonnet, Noe, River, and Sebastian

Chapter 1. *Design with Nature*: Promises and Pitfalls

> *In order to endure we must maintain the great cornucopia which is our inheritance. It is clear that we must look deeper to the values which we hold. These must be transformed if we are to reap the bounty and create that fine visage for the home of the brave and the land of the free.*[1]
>
> —Ian. L. McHarg, *Design with Nature*, p.5

I knew little of ecological complexity while attending architecture school in Miami in the early 1970s. I knew less about the Florida Everglades, the "River of Grass" so dubbed by writer-environmentalist Marjory Stoneman Douglas in an effort to mitigate the insidiously ignorant reference to the place as a "swamp." And yet there I was, standing along the side of Alligator Alley, the two-lane road between Miami and Naples that cuts through the heart of the Everglades, admiring the emergent sawgrass for as far as the eye could see.

The occasion was the study of Cheekees, the airy thatched structures that have sheltered the goods and people of the Miccosukee tribe for centuries. Many of them remain in use as part of a small roadside village conveniently close to the Miami city limits. Some serve as a marker for flat-bottom boats, watercraft that can traverse standing water or emergent reeds and sawgrass equally well. Farther into the Everglades, dome-shaped stands of cypress trees can be seen dotting the watery expanse, capturing in their verticality the ever-shifting play between bright light and passing shade. It is a mesmerizing landscape, less for any single standout feature than for the measured pace of its diurnal and seasonal change (see Figure 1.1).

About the time I turned in for review a meticulously scaled replica of a Cheekee, another expansive grassland became etched in my mind, this time via a documentary, *Multiply . . . and Subdue the Earth*. The on-screen image showed a vast Midwestern meadow with nary a vertical counterpoint save for a tall, mustachioed man standing resolute amid the vegetation: Ian L.

1

Reconsidering Ian McHarg

Figure 1.1. A cypress dome rises out of the "River of Grass" in the Florida Everglades. The trees take root in shallow solution holes that over time accumulate layers of organic material, deeper at the center and shallower at the edges.

McHarg. Many people became acquainted with McHarg through this documentary. It was, for me, a momentous introduction. I cannot recall the details of the script, but the message through McHarg's voice was clear: the time was here and now to protect the earth and guide humankind to a shared, mutually beneficial existence with nature.

I had to be part of it. What little I had seen of the Everglades made it so. I then rushed to the campus bookstore and purchased McHarg's book, *Design with Nature*, which bore an endorsement on the back cover, calling it "the most important book of the century."

One can only guess what must have crossed McHarg's mind a few years later when I unfolded a design portfolio before him in the hope of gaining admittance to the master of landscape architecture program he chaired at the University of Pennsylvania. Following undergraduate studies I had found work in a small Miami landscape architecture office specializing in planning single-family subdivisions—many of them showcasing cookie-cutter homes spaced along curving roadways and cul-de-sacs set comfortably over excavated Everglades marl and crushed oolite limestone. Plans of this development lay open on his desk, including planting schemes

Design with Nature: Promises and Pitfalls

festooned with rubber-stamped trees. Absent in the presentation was any recognition of an underlying ecology or sensitivity to the bygone South Florida landscape, an omission that surely helped intensify McHarg's redemptive educational mission. He offered the eager presenter a scholarship on the spot, contingent on first completing introductory courses in biology and earth sciences.

Following this coursework, I returned to the Everglades and was able for the first time to truly marvel at its water-powered ecology: majestic nimbus clouds, deep into the landscape, with feathery anvils 60,000 feet up in the air and with furious, dark underbellies lashing at the sawgrass below, literal gushers of life to myriad creatures large and small, from the American alligator to the Florida apple snail. As hydrology goes, few places on earth offer evidence of the daily exchange between land and sky at such a grand and palpable scale. There is inherent beauty in nature's work, all the more so if one knows how and why it is so.

This, above all else, was McHarg's pedagogical intent: to teach impressionable young people—architects especially—how nature works and, in doing so, fortify their commitment to and skill at designing with it. This was the clarion sounded by *Design with Nature*. Enhanced by McHarg's fervent writing style, the book set the stage for a magnificent indoctrination, one that begat many apostles of the faith, including me.

And so, many years later, it surprised no one gathered to interview consultants for the preparation of a master plan for the South Livermore Valley in California when this author and his partner, Stephen Hammond, summoned the promise of *Design with Nature* as the best way to develop the valley while preserving and even enhancing its natural and rural beauty. It was 1997 and we had at our disposal the best possible precedent with which to support our claim: the *Plan for the Valleys*,[2] prepared three decades earlier by McHarg and his partner David Wallace. Echoes of this celebrated project filled the conference room as landowners, community representatives, and City of South Livermore and Alameda County planning officials listened to WRT's statements of approach and qualifications. (More on the South Livermore Specific Plan in Chapter 3).

The key to the pitch was "fitness," the idea that the proper analysis of environmental factors—geology, hydrology, soils, vegetation, wildlife,

3

among others—can bring to light a landscape's intrinsic suitability for a given land use program, be it development or conservation.

Ecology as a Promised Land

Fitness is the ultimate aim of *Design with Nature*, the path toward environmental health and beauty. McHarg based the truth of this thesis on the multifarious ways in which organisms adapt in the struggle for survival and the resulting correlation between natural form and ecological function. This was the lesson he hammered home year after year to the graduate students who came from across the globe to study under his wing. To McHarg, nature's own bent towards fitness was a creative act:

> Evolution consists of a tendency towards increasing fitness whereby the organism adapts the environment to make it more fitting and, through mutation and natural selection, adapts itself towards the same end. As the process of fitting exhibits the direction from simplicity to complexit . . . it corresponds to the most creative processes on earth. Processes whereby the system reverts from complexity to simplicity . . . are therefore entropic and destructive. These are two polar conditions, the first creative fitting and the other destructive and unfitting. The measure of fitness and fitting is evolutionary survival, success of the species or ecosystems, and, in the short run, health.[3]

These are among the most significant words in *Design with Nature*—its very promise. They underscore the view that only the mutually beneficial adaptation between humans and nature—between a development program and a site's underlying ecology—can be considered fit and, therefore, creative.

Listening to McHarg speak about fitness could be a revelatory experience. Three vital arguments spewed forth. First, that the process of planning and design was deterministic, one that was based on factual information leading to defensible conclusions. If there was nothing arbitrary about nature's work, then neither should there be in the work of man (so long as ecological health was the overriding goal). Second, that the concept of ecological health applied seamlessly to all scales of inquiry, from the region to the specific site. Accordingly, the artifice of property lines and political boundaries was meaningless, a patently un-ecological way to view the world perpetuated by misguided bureaucrats, mapmakers, and school teachers the world over. And third, that addressing fitness required the expertise of multiple disciplines—of designers, planners, and scientists alike, the latter providing the requisite exactitude in pursuit of fact and defensibility. These were the attributes that made McHarg's vision transcendent: it was not merely a way to practice a profession but also a way to change the world, site by site, region by region, in lock-step with others.

However, McHarg neatly divided the world between that which was fit and that which was not. Nature was fit, cities were unfit. He regarded cities as "scabrous" entities: grimy, scummy, vermin infested, and beset by disease—a Dantesque form of earthly imprisonment. Few cities exemplified this ugly predicament more than his hometown of Glasgow, Scotland. He lamented the city's careless growth and its impact on the wildlife he cherished as a young lad:

> Lark and curlew, grouse and thrush had gone, the caged canary and the budgerigar their mere replacements. No more fox and badger, squirrel and stoat, weasel and hedgehog but now only cat and dog, rats and mice, lice and fleas.[4]

THE LAYER CAKE

McHarg's views of cities are succinctly conveyed by the cover of the first edition of *Design with Nature*: a dark, foreboding skyline, diminished by the weight of its own fowl air and grime. It is in the perceived stark contrast between city and nature where the soul of *Design with Nature* resides. He spared no word, no critique, in advocating nature as the source of health and well-being—a paradisiacal ecology. The challenge was figuring out how to inhabit this ecology without ruining it; in other words, how to live in paradise. The solution was the "layer cake," the method by which scores of students learned how to decode a site's ecology and assess its suitability for urban development.

A precursor to the geographic information systems that support every modern-day planning practice, the Layer Cake was built of strata representing individual natural factors such as bedrock geology, aquifer recharge areas, slopes, soils, vegetation, and the like. From this representation, one assessed where, say, homes or a highway could best be located without upsetting the natural ecology. Layers of information were gathered from a variety of sources, including the U.S. Geological Survey, U.S. Department of Agriculture, and aerial photography. Prior to the advent of computer-based mapping, a site's ecological attributes had to be graphically superimposed by means of color markers applied on transparent sheets of mylar, a tedious exercise that nonetheless could produce astoundingly beautiful results.

Figure 1.2. The Valleys are approximately 10 miles NNW of Downtown Baltimore; the Baltimore Beltway, I-695, is in the foreground. The Plan envisioned single-family home clusters tucked in the forested slopes and plateaus framing the valleys, which are valuable aquifer recharge areas.

THE PLAN FOR THE VALLEYS

In the early 1960s McHarg applied the "layer cake" to the planning of the Green Spring, Worthington, and Caves Valleys near Baltimore. Known as *The Plan for the Valleys*, the project area comprised approximately 14,000 acres outside the Baltimore Beltway (I-695), more than ten miles from the city's downtown. In this pioneering effort, the layering of ecological data pointed to the preservation of the flatlands in favor of development on the valley walls or the forested as well as unforested plateaus—a counter-intuitive result. More than 26,000 dwellings were proposed, distributed in residential clusters, hamlets, and village centers. Parks, greenways, and natural areas flowed through the development zones, constituting in open land and community facilities nearly one-third of the total project area (see Figure 1.2).

Economists retained by Wallace and McHarg projected a $7 million greater return on the sale of land based on the proposed conservation plan rather than under an uncontrolled growth scenario. The plan thus demonstrated, at least on paper, that ecologically-informed development could be both environmentally sound and profitable. *The Plan for the Valleys* was "fit" and "creative"—a new and powerful way to direct development toward and within exurbia. Rather than expunging nature, *The Plan for the Valleys* let development coexist with it. To McHarg, this seminal achievement pointed to a redefinition of the American landscape, fortified by the promise of a fulfilled populace:

> The United States awaits a large-scale demonstration of a beautiful landscape developed with wisdom, skill and taste, the evolution of a process which can produce a noble and ennobling physical environment, a step towards the American Dream.[5]

Except that it never happened. Wallace and McHarg had argued that a private real estate syndicate and conservation trust were needed to manage the open land and equitably distribute the profits from land sales among the 250 landowners who had signed on for the project. This did not come to pass, and without such a legal foundation the implementation of the plan languished. It also did not help that a series of economic recessions in the 1970s effectively arrested the expansion of urban development into Baltimore's prized rural landscape. Without growth pressures, why fuss over a complicated and contentious legal implementation mechanism? An update to *The Plan for the Valleys* was prepared by WRT in 1989, but it too failed to institutionalize a transfer of development rights. In his memoir, David Wallace rendered the verdict:

> The County, the Baltimore Regional Planning Council, and the State of Maryland all adopted the plan 'in principle'. Damaging zoning changes and inappropriate utility and highway layouts [were] perverted, or coerced into conformance. But unfortunately the Valleys cannot be considered, as has been touted, to be the first successful large-scale example in America of humane development and conservation of the countryside by citizen action.[6]

William Roberts, McHarg's long-time partner and an esteemed landscape architect in his own right, asserts that the *Plan* "at least protected valued scenery, although to the benefit of a social elite and for no better reason than the projected growth in Baltimore County did not materialize" (author's recollection of a conversation with Roberts on the subject). By "social elite" Roberts is alluding to the wealthy homeowners who prize the valleys' horse-breeding and fox-hunting setting. The *Plan*'s exclusionary motive was emblazoned above the report's preamble by a hand-drawn vista

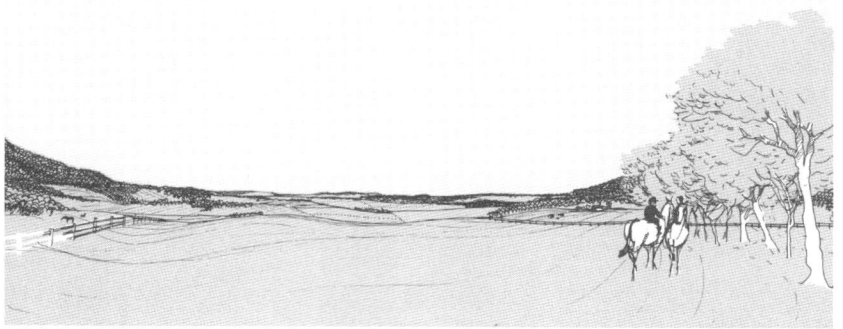

Figure 1.3. Owing to the absence of a transfer of development rights, Wallace's economic projections, attached to McHarg's ecological planning method, were never tested. The Valleys today have not appreciably changed since the plan, except for strip commercial development along adjoining highways.

of one of the valleys, with two horsemen admiring the countryside, devoid of development save for a distant residential manor (see figure 1.3). A drive through the valleys today reveals nothing more than scattered homes wedged in the woods or poking out into open fields, a low-density pattern of development at the very low end of the density spectrum.

Still, *The Plan for the Valleys* remains a monument to the vision McHarg professed in *Design with Nature*. It certified that through a rational planning method, communities could well grow outside the grime of cities while preserving the intrinsic ecological and social value of the exurban landscape. To landowners and developers operating in the rural milieu, this was and still is sweet music, exaltation to the idea that home buyers intent on escaping the city can have a home amid a preserved natural or rural landscape.

Many firms have profited from the planning and design of "environmentally correct" suburban development in greenfield sites. As with WRT, many have been led by graduates of McHarg's landscape architecture program at the University of Pennsylvania. Some, like Rahenkamp Sacks Wells Associates, made suburban planning virtually their sole brand. I worked part-time at RSWA while attending Penn. Hours upon hours were spent constructing maps of environmental factors justifying the layout of new homes in rural New Jersey and elsewhere. Much of this work was done in support of legal arguments intended to secure development permits. Land use attorneys were often consulted about how best to represent the land's intrinsic fitness for development. Rendering environmental factors in graphic layers for judges and lawyers to decipher was a recurrent challenge. The "environmental composite" graphic became a tool of persuasion, a way to legally record the avoidance or mitigation of environmental impact relative to the layout of roadways and home sites. Economic analyses were conducted to confirm the optimum development intensity and mix of uses and building types. Finally, a development plan was drafted closely matching the regulatory bounds. The result: idyllic communities in the hinterland, miles from urban centers.

The rural milieu, of course, would not have existed as developable land if not for the federal subsidy of interstate highways, which brought vast regions of open land within reasonable commuting distance of urban employment centers. The growth pressures that precipitated *The Plan for the Valleys* would not have materialized had it not been for three interstate highways that bound the place: I-695, I-795, and I-83. It would have been well within McHarg's ecological acumen to condemn the extension of highways into pristine areas. Cleary, the best way to protect a sensitive landscape is to restrict access to it. However, configuring environmentally-acceptable highways is given prominent mention in *Design with Nature*. The book goes into great depth in demonstrating how the "layer cake" could reveal the most advantageous alignment of a five-mile stretch of the Richmond Parkway in southern Staten Island. The same method was applied to a section of Interstate 95 in New Jersey between Philadelphia and Trenton. Regarding such road-building enterprises, McHarg wrote:

> Within limits set by points of origin and destinations, responsive to physiographic obstructions and the pressure of social values, the highway can be used as a conscious public policy to create new and productive land uses...[7]

Through the ecological planning method, McHarg was certain that highways could be better designed and built, facilitating a more rational regional pattern of development. Given current concerns about carbon emissions, however, it is difficult to accept that *Design with Nature* prescribed an environmentally, economically, and socially sound way to increase their output.

Abetting Sprawl, Failing the Inner City

Since the advent of the National Environmental Policy Act in 1969, thousands of miles of highways have been constructed nationwide in accordance with environmental review processes that generally follow McHarg's pioneering methods for the evaluation of ecological, economic, and social impact. More important, ecological planning as a basis for environmental review has been widely used in the adjudication of urban development upon lands rendered accessible by federal and state highways. Such growth overwhelmingly benefitted the white middle class, facilitating their flight from the inner city and leaving in their wake a disadvantaged class—African Americans mostly—in blighted urban conditions.

Having the white middle class leave the city for greener suburbs and commute to and from work along fresh new highways was the predictable outcome of federal transportation, economic development, and even national defense policies. To a majority of Americans it was the rational thing to do, especially as a response to perceived inner city crime and social upheaval.

Caused by a chronic lack of affordable housing, economic inequalities, and police brutality, tensions exploded within inner city African American enclaves soon after the enactment of the Civil Rights Act of 1964. Triggered by the arrest of a black motorist, the 1965 Watts riots in Los Angeles left 34 people dead and more than a thousand injured. In 1967, the unfair arrest of a black cab driver in Newark, New Jersey, caused public confrontations with police that left 26 people dead and more than 700 injured. A scant week later, a Detroit police raid on a local African American hangout sparked a riot that over five days left 43 people dead and more than 1,100 injured.

These and similar events throughout the nation crystallized the plight of African American communities, fueling black militancy. The assassination of Dr. Martin Luther King in April 1968 triggered another wave of riots in virtually every major U.S. city, adding further impetus to the abandonment of urban core areas by the white middle class. Many commuting highways hovered over the poverty below, having wiped out swaths of well-established African American neighborhoods through the exercise of eminent domain. For example, the extension of a section of I-95 north of downtown Miami effectively destroyed the historic Overtown community in 1968, displacing 10,000 people from their homes.[8] In Washington, D.C., Interstate 295 was completed a year earlier, at the edge of the historic African American community of Anacostia, effectively separating it from the Anacostia River—and the rest of the city. The highway is a noisy, carbon-spewing barrier to the low-income hillside residents, but a speedy conduit in and out of the city for the suburban residents of Prince George's County, many of them commuting government workers.

Earth Day was instituted in April 1969, an apogean event in the course of the environmental movement. McHarg was a leading figure in Philadelphia's inaugural event (reputedly the nation's largest), which included the participation of Edmund Muskie, Ralph Nader, and Alan Ginsberg among other environmental advocates. It was celebrated barely a year after the nation was beset by angry black mobs rightfully protesting the state of their lot. How many miles of highways have been built since then? How many square miles of development laid out on pastoral land? How many millions have chosen suburban over urban life in the name of personal safety and health? How is it possible that the Civil Rights Act proclaimed racial desegregation while the government at the same time was sanctioning, through a massive highway building program, the residential separation between blacks and whites? Under such backdrop, to inner city African Americans the concept of "design with nature" would have been at best incomprehensible, at worst regarded as yet another instrument for their own suppression by the privileged class.

McHarg was well convinced that the understanding of ecological process had as much relevance to the planning of cities as it did to the hinterland. Several case studies of urban land are included in *Design with Nature*. He explained at length the underlying geology and hydrology of the nation's capital, for example, as well as the physiography of the Piedmont and Coastal Plain and how the seam between them influenced the location of major federal buildings, such as the U.S. Capitol. Missing, however, are references to the condition of the urban environment and its impact upon affected communities. Not one word, for example, about the Anacostia River and its abysmal water quality, the result of combined sewer outfalls and overland pollution. Or about the shaping of the river by the U.S. Army Corps of Engineers and the ensuing environmental degradation which, along with toxic landfills and soot-spewing power plants on its southern shore, further isolated the impoverished Anacostia African American hillside communities.

This omission is surprising given McHarg's high regard of Olmsted's legacy, especially in light of Frederick Law Olmsted Jr.'s work in Washington D.C. As a member of the McMillan commission, Olmsted Jr. promoted the creation of a water-oriented park in the upper reaches of the Anacostia River, to include water-cleansing wetlands. Olmsted Jr., in effect, was attempting to follow his father's success in the fens of Boston.

And yet, in the context of the inner city, the idea of a "working nature" is absent in *Design with Nature*. McHarg's master theory was focused, rather, on the salutary effects of nature well beyond established urban areas (as exemplified by the *Plan for the Valleys*). Out of focus was the nature of cities, especially those associated with poor, racially segregated communities. As discussed in Chapter 4, it would take 15 years following the publication of *Design with Nature* for the idea of an urban ecology to become prominently discussed within the landscape architecture profession—by Anne Spirn, McHarg's student and his successor at Penn. And even then, this laudable effort was but a whimper in comparison to the environmentally sanctioned production of suburbia, which continued to roar through the twentieth century up to the collapse of the housing market in 2010.

It would be wrong to attribute the relative abandonment of core urban areas to *Design with Nature*. Still, McHarg's thesis abetted white flight from cities, delivering both the ethical foundation and technical means by which a "scabrous" urbanity could be left behind in favor of an "idyllic" lifestyle amid a natural setting.

Design with *Which* Nature?

McHarg's pedagogy toward ecological design ultimately rested in the notion of a natural realm bigger than humankind, one of immense richness and answering in its intractable complexity to the divine. Humans are on earth, he would argue, to be guardians of this realm: "What better concern can there be than nature? What better role than to be a modest steward?"[9] (McHarg's deification of nature tilted toward pantheism, not to "an ultimate God, but to lesser yet probably palpable manifestations," by which he meant Gaia.)

McHarg believed that, to exercise such stewardship, nature's richness and complexity had to be understood, its sustaining power unlocked and managed with due cognition of its governing laws. *Design with Nature* illustrates the point through a fictional astronaut who strives to understand the interrelationship and value of finite natural resources in pursuit of a self-sustaining eco-system (i.e., his own abode in outer space). I was so taken by this science fictional approach to ecology that while at Penn I lobbied for and received McHarg's permission to take an elective course on space exploration. The course was focused on space habitation technologies and the means to harvest the moon, Mars, and other planets in support of orbital life. The textbook was Gerard K. O'Neil's *The High Frontier: Human Colonies in Space*.[10] Assessing the suitability of any sidereal abode—especially ours—begs the question of nature, its value, and the degree to which it should be managed. Three distinct scenarios come to mind.

Scenario One magnifies the course we are on: the relentless exploitation of natural resources leading to a transformed planet with an altered landscape and changed climate. Polar ice caps melt, extreme weather intensifies, lowlands everywhere become submerged, and the planet convulses through increased tectonic activity. In this scenario, humanity accepts the extinction of many animal and plant species, reducing them either to zoological and botanical displays or samples in gene banks for the benefit of scientific research. As a corollary, animals and plants that help support human nourishment, physical health and psychological comfort are preserved, domesticated or genetically improved according to the needs and desires at hand. By the standard of biocentrism—the belief that species have rights of fair treatment and should be allowed to thrive and perish without human interference—this scenario represents doomsday. On the other hand, to those who accept the Biblical call for humans to "dominate and subdue the Earth," the domestication of nature is not a bad result at all; it is simply the "natural" thing to occur. Such basic denial of the humanity's responsibility toward the well-being of the mother ship inspired McHarg to fight against it to the core of his being, in the process defining his mission as teacher, lecturer, and practitioner.

Scenario Two might be the antithesis of the first, the total preservation of nature. Here, humanity is struck by a deep love of all biota, its infinite complexity and life-giving beauty. Overcome with collective guilt over the planet's diminished ecology, world leaders embark on a millennial restoration project: the transport of every human being to an orbital world, allowing animals and plants to recolonize the Earth and adapt anew in a depopulated environment. As a prize, people high above marvel at the wild spectacle below through remotely-gathered imagery transmitted in real time to every living pod. Precisely managed high-tech ecology within each pod produces the requisite water, air, food, and waste treatment. Every water, air, and organic molecule is valued as a shared vital resource. Astro-ecologists reign supreme, governing over every aspect of civic and political life. Individuality is eradicated from the natural resource equation, transferred instead to the realm of the arts, especially the digital pixilation of the virtual facades that define living and working quarters. Earth Day, April 22, becomes that start of the calendar year, supplanting the old timekeeping system.

Scenario Three would be the middle ground between the first two. Cities, farmland and wilderness areas become distinctly defined and geographically separated from each other. Regions of wild lands in every biome are accorded natural conservation and regeneration status. Within them only dispersed towns and villages are permitted, mainly as loci for ecotourism and scientific research. Initiatives such as the Buffalo Commons become a reality, with a vast swath of land from Montana to Texas reverting to shortgrass prairie and inviting herds of buffalo to once again roam freely in impressive numbers. Similarly, the Comprehensive Everglades Restoration Plan acquires a new lease on life, with federal funds appropriated under enthusiastic bipartisan congressional cooperation. The channelized Kissimmee River is permitted to reestablish itself within a natural floodplain, dikes are removed from Lake Okeechobee, and the sawgrass-powered water cycle is restored to the benefit of the roseate spoonbill, snail kite, and Cape Sable seaside sparrow. Under this scenario urban development is restricted to preexisting urbanized areas: densities rise but are mitigated by green infrastructure; access to recreation is enhanced through expanded greenways and parks; and high efficiency mobility and building design dramatically reduce the use of non-renewable resources. Global warming finally slows.

Each of these scenarios represents well-known ethical positions with respect to humankind's relationship with nature. Had space shuttles and an international space station existed in his time, under Scenario Two John Muir could well have applauded the extrication of humans from the planet if it meant preserving the Earth as God's temple. Under Scenario Three Thoreau and Emerson could well have endorsed strict urban limits had this promoted regionally managed wilderness areas available for solitary immersion, contemplation, and rejuvenation. There are plentiful standard bearers for the subjugation of nature as posited by Scenario One, but William Mulholland deserves the nod. As superintendent of the Los Angeles Water Department in the late 1890s, he masterminded the diversion of water from Owens Lake, a natural reservoir fed by the snowmelt of the Sierra Nevada, to the San Fernando Valley. He succeeded in irrigating not only crops but also the advent of urban sprawl as a hallmark of the Los Angeles basin.

Not one to dwell in fantasy or doom, McHarg's ethical framework clearly favored the middle ground: a designed and managed urban world in coexistence with nature—enough of it, at least, to guide the human spirit toward transcendent naturalism. But the odds have never been in his favor. The allure of the Buffalo Commons or a reconstituted Everglades notwithstanding, the chances of whole towns or agribusinesses being eradicated from vast regions of the United States for the purpose of creating, in essence, large-scale zoos are not very good. And the odds of a halo-like orbital world above a pristine planet are strictly science fiction.

We are left, then, to contend with the first scenario: a largely urbanized world with scant evidence of a paradisiacal ecology. We might consider Bill McKibben's *The End of Nature*[11] the emblem of such a future. Published in 1989, the book paints a sobering picture of the state of the world every bit as disturbing as Rachel Carson's *Silent Spring* did a little more than a quarter century earlier. But, where Carson was focused on the preservation of a nature as an embattled paradise, McKibben treats nature as a paradise already lost.

Other critics posit that nature is nothing more than an artifice of our own creation and that, within such construct, "nature" is but a simulation of the real (but vanquished) thing.[12] After all, how natural can a place like the Florida Everglades be when human-controlled dams, levees, floodgates, and canals regulate the amount of water that those majestic nimbus clouds shed into the sawgrass below? How pristine can it be when the Burmese Python, a human implant, is wreaking havoc upon the native fauna with no solution in sight for achieving its eradication? What is nature, then, if one of our most revered wilderness parks is, in the end, a landscape utterly affected by human agency? How do we reconcile our inclinations toward ecological restoration, as in the Everglades Plan, with having to contain such efforts within negotiated political boundaries?

Although it is not the purpose of this book to settle the question of what nature is or is no longer, it is nonetheless relevant to discuss the extent to which "what's out there," beyond the reach of cities and farmland, is of import to human health. To McHarg, the value of "what's out there" resided in the salutary effects of a restorative realm at the low end of the human interference spectrum—a "natural" place to live in. And therein lies the ultimate pitfall of *Design with Nature*, for if this realm is no longer pristine, if it is, in the end, something of our own creation, what nature is there to really design with? Moreover, if we accept that humanity is destined to live in urbanized land, then we must ask: what is an optimum *urban nature* and how should we design with and within it?

Chapter 2. The American Wilderness:
An Evanescing Myth

> *If you travel much in the wilder sections of our country, sooner or later you are likely to meet the sign of the flying goose—the emblem of the National Wildlife Refuges. Wherever you meet this sign, respect it. It means that the land behind the sign has been dedicated by the American people to preserving, for themselves and their children, as much of our native wildlife as can be retained along with our modern civilization.*
> —Rachel Carson, 1907–1964

During clear springtime days, common snapping turtles at Philadelphia's John Heinz National Wildlife Refuge like to emerge from their marshy digs, climb onto low-lying tree limbs, rocks, or dry patches of mud and bake there, motionless for hours. It can be unnerving to watch them from a nearby boardwalk, much like playing a game of blink. Who will move first, the casual visitor twitching with impatience, or the slow and steady turtles, oblivious to the blinking eyes? Alas, it is never a contest. If a time-lapse video of the place were taken and played back, it would show the turtles perfectly still while people move in a blur.

Across the open wetland there is a small cove ringed by bull rushes, a reliable place to spot a great blue heron. The bird's wingspan is awe-inspiring, its glide over the water graceful and effortless. Inland there is a well-placed bench at the far reaches of a long meadow where either summoned patience or chance will reward the bird-watcher. A bald eagle may show up and hover in circles above, a flock of wild turkeys will peer out from the wooded edge, or a Great Horned Owl will fly across the field, punching through the trees so suddenly it will draw a gasp. Along the gravelly trail that encircles the marsh an eastern garter snake will coil in defiance if a hiker approaches distractedly or mistakes the reptile for a fallen twig. Deer, predictably, occupy the preserve as well, although it takes a keen eye to spot them in the dappling understory.

In pre-colonial times the refuge, formerly the Tinicum Marshlands, would have been part of a nine-square-mile drainage delta comprising woodland and freshwater marshes. Today, only 1,200 acres remain, 200 acres of which account for the largest remaining freshwater marshes in all of Pennsylvania. Still, the place provides a genuine escape from the bustling city—almost. The southern end of the refuge is a mere half-mile from Philadelphia's International Airport. With a flight departing every half-minute on average, one is far more likely to catch sight of an ascending airplane, engines revving, than to witness the dive of a red-tailed hawk or the flick of an indigo bunting. The ever-present din from nearby Interstate 95 and scheduled rumble of freight and passenger trains along adjacent rail lines distract even the most serene birder. But it is not only external distractions that preclude visions of Walden Pond: a 20-foot-wide swath cuts through the floodplain woodland, bearing markers of the oil pipeline that runs beneath. Surely few other places in the United States exhibit such cheek to jowl coexistence between a protected wildlife and behemoth infrastructure (see figure 2.1). (The Tinicum Refuge was established in 1972

Figure 2.1. Six miles from downtown Philadelphia, the John Heinz National Wildlife Refuge is surrounded by industry, rail lines, highways, and the Philadelphia International Airport. It is, however, cheished as a "natural oasis."

through congressional legislation. Late in 1991 the name was changed in memory of Pennsylvania Senator John Heinz, who had fought to preserve the marshlands. He had perished earlier that year in a tragic airplane crash over a lower-school in adjacent Montgomery County.)

WILD LANDS THAT HARDLY WERE

National wildlife refuges are managed by the U.S. Fish and Wildlife Service (FWS). There are 520 discrete reserves, or "units," totaling 93 million acres within the 50 states, plus the territories of American Samoa, Puerto Rico, the Virgin Islands, the Johnson Atoll, Midway Atoll, and several other Pacific Islands. These reserves are home to more than 700 species of birds, 220 species of mammals, 250 reptile and amphibian species, and more than 200 species of fish. Fifty-nine refuges have been established with the primary purpose of conserving threatened or endangered species. Overall, 280 of the 1,200-plus federally-listed threatened or endangered species in the U.S. are found within them. And yet the refuges represent far more than habitat for valued wildlife. They embody "*a priceless gift—the heritage of a wild America that was, and is*" proclaims the mission statement of the FWS. Altogether, the refuges represent about 3.8 percent of the total land area of the United States. Our nation's creation myth—that a bountiful and boundless terrain over which to roam free in search of opportunity—is unprotected but for this modest collective parcel.

In *The Day Before America*[1] William MacLeish describes the relationship between Native Americans and the land, lending credence to the notion that early European settlers encountered something other than a pristine continent to civilize and call their own. The burning of woodland was already a common landscape management practice. From the clearings, maize was grown as a staple and, in the case of the Iroquois, used as a valuable trading commodity. Sumpweed, goosefoot, and sunflowers were also cultivated. In the Rockies, sheep and goats were domesticated, while in the Great Plains, the prairie was burned and the bison driven to their deaths at designated cliffs to sustain the native tribes. In the Southwest, the Hohokam built miles of canals to irrigate thousands of acres for the cultivation of cotton, tobacco, maize, beans, and squash. Imagining himself in pre-colonial times, MacLeish writes:

We cleared the way for plants that would be our allies. Our fires burned woodland and prairies, opening canopies for sunlight, creating mosaics and edges where ruderals could flourish. We tore the earth in our camps, wearing away the cover with our feet, digging in it with sticks to make dumps for our wastes. And the weeds came in, drawn by the disturbance and the concentrations of nutrients we provided.[2]

Where McLeish relied on the archaeological record to construct his tale, Álvar Nuñez Cabeza de Vaca's was based on the memory of an extraordinary personal trek. He was the treasurer of a Spanish expedition that shipwrecked in 1527 on the Florida coast near present day Tampa Bay. A handful of men survived and returned home, but not before wandering for eight years along the Gulf Coast and Rio Grande toward Mexico, suffering first-hand the customs and hardships of native peoples who had barely begun to tangle with Europeans. Upon his return to Spain, de Vaca wrote Relato, the very first account of the people and landscape of the New World. Of life on the Island of "Ill Fate," known today as Galveston, Texas, he recounts:

The Indians of the interior protect themselves in another way that is even less bearable: they walk around with firebrands in their hands, burning the fields and the woods around them to drive off the mosquitoes and to drive out from under the ground lizards and other things they eat. They also kill deer by encircling them with fire. They also do this to destroy the animals' grazing areas, so that they will be forced to go where they want them . . .[3]

There is little question that when William Bradford and his fellow Pilgrims landed on Cape Cod in 1629, America was already a human-altered landscape. Native Americans had long adapted the land to their own use. Still, as Roderick Nash argues in *Wilderness and the American Mind*[4], the image of a boundless wilderness took root in the American psyche from the very beginning of colonial times. This is understandable. To the European settlers accustomed to a cultivated landscape, advanced animal husbandry,

The American Wilderness: An Evanescing Myth

and refined methods of social exchange, the American landscape must have indeed appeared as a vast wilderness—both natural and cultural. The zenith of the representation of the American landscape as wilderness occurred in the mid-to late-nineteenth century by a group of painters who portrayed the more remote and imposing features of the landscape with luminous brushstrokes tethered to the sublime. Arguably, the Hudson River School's most capable exponent was Albert Bierstadt. In 1865 he produced *Looking Down Yosemite Valley*, a massive five-by-eight-foot panel presenting a view of the majestic landscape deeply embedded within the Sierra Nevada. Only a year earlier, Abraham Lincoln had signed into law its preservation as a public resource, the first such instance anywhere in the world where a wilderness area was so designated.

Lincoln's action set the stage seven years later for the creation of Yellowstone as the United States' first national park. A genuine American invention, the designation of scenic wild landscapes as "national treasures" underscores the fundamental attachment of the American psyche to the idea of a *natura immaculata*, a pure, unblemished nature. With travel to far-off natural destinations limited for most Americans, the paintings of the Hudson River School stood before the general public as evidence of the nation's majestic (if not quite utterly pristine) landscape.

Few national park visitors stop to ponder the true extent of our natural heritage or the great distances that separate the remnants by which it is embodied. National parks and wildlife refuges are but dots on the map, overwhelmed in size by metropolitan agglomerations in the east and military reservations in the west. The 3,200-acre White Sands Missile Range in New Mexico, for example, is nearly 50 percent larger in area than the Grand Canyon, Brice Canyon, and Painted Desert National Parks in Arizona, combined. The FWS refuges fare no better. Supawna Meadows National Wildlife Refuge, the nearest to the Tinicum marshlands, lies 22 miles to the southwest in Salem County, New Jersey, next to a landfill. Only nine other refuges are located within a 100-mile radius of Philadelphia, all in neighboring states. The next closest in Pennsylvania is the Erie National Wildlife Refuge, about 300 miles from Philadelphia at the opposite end of the state. Most resource-based national parks and wildlife refuges exist in remote locations. Long journeys are necessary to reach them, a form of pilgrimage that fits well with the nation's early Christian, pioneering heritage. But when pitted against farmland, urbanized areas, and the infrastructure that binds them, natural preserves as a combined territory are fairly inconsequential.

Infrastructure Rising

Infrastructural development has molded the conception of the American landscape as much as or more so than the collection of preserved wilderness areas. How we see and move about the land, what we hold dear as a shared landscape, has greatly been shaped by major waves of infrastructure. The first such wave was the Public Land Survey System (PLSS), established by Thomas Jefferson in 1785.

The PLSS had a dual purpose: to transfer federal lands into private hands as a way to replenish the nation's coffers after the Revolutionary War and to better manage the expansion of a national agrarian economy. Land development was thus parsed, giving shape to roads, towns, and farmland as settlement crept westward. Flying across the country, the orthogonal geometry of townships and sections established by the PLSS is relentless, with county roads often following survey base lines and meridians far into the distance. Rivers, streams corridors, and hillocks puncture or bend the grid, seemingly struggling to break free but seldom failing to become reinscribed again and again, yielding from high above a rectilinear mosaic of a captured nature. Against mountains, the PLSS lines will disappear, only to reemerge, perfectly aligned on the other side as if they had bored a hole straight through. Not even the mighty Colorado River manages to nudge the grid-imposed border between Utah and Arizona.

The "national grid" has entered our collective consciousness through myriad roadways that trace section lines across the landscape. How many films portray roadways in the middle of nowhere stretching for miles into nothingness? Or a lone character appearing at a desolate rural crossroads, each dusty lane extending straight and far into the distance? John Ford's *The Grapes of Wrath* opens with such a scene as Tom Joad walks with determination towards his boyhood home. That the home is nowhere in sight is an omen of its ultimate abandonment. The ensuing departure to California from dust-ridden Oklahoma follows U.S. Route 66. The highway of lore

The American Wilderness: An Evanescing Myth

is drawn as the classic evanescing line through the arid southwest landscape, a metaphor of the arduous journey and uncertain destination faced by the migrating Okies. In reality the prize for a stretch of pavement goes to Route 46 in North Dakota. With a length of 123 miles, it qualifies as the longest, straightest road in the nation. Three towns dot the highway, Gakle, population 335; Enderlin, population 947; and Kindred, population 614. An approximation of such sparseness was given cinematic resonance in *Badlands*, the acclaimed 1973 Terrence Malick portrait of troubled lives in South Dakota that go off society's gridded path.

The second major wave of infrastructural development belongs to the railroads, standard bearers of nineteenth century industrial power and grime. In the United States a milestone was achieved on May 10, 1869 in Promontory Summit, Utah, when tracks laid separately from the east and west were joined to form a continuous rail line between Omaha and Sacramento. Railroad baron Leland Stanford did the memorializing honors by driving the final spike, a golden one. Built by the Union Pacific and Central Pacific rail companies and backed by federal funding and land grants, the Transcontinental Railroad united the nation with a network of railroads, coast-to-coast. It was the engineering feat of its era, much as sending a man to the moon and back would be a century later. (That each enterprise took six years from the time Lincoln and Kennedy, respectively, proclaimed their national importance—presidents who would ultimately be assassinated—cannot escape attention.)

Today there are 223,000 miles of rail line in the United States, enough to go nine times around the equator. On a nationwide map, the tracks seem to run freely across the land, clearly at odds with the PLSS grid. The railroads simply had to follow the dictates of the land, across mountain passes or along river banks where the grade is gentler, a proto "design with nature" accommodation. But it is where fitting land stops that railroads contribute to our nation-building lore: over deep ravines, across wide rivers, and through established urban centers where bridges and trestles hold gravity at bay. They rise over the ground in commanding contrast to anything around them, yet they somehow fit in, anchoring the landscape as well as our collective imagination. Many such structures have achieved significance as engineering feats, others as cultural artifacts.

In Dallas the Union Pacific rail line was built at the edge of the city's downtown, paralleling the Trinity River floodplain. After the great flood of 1908, levees were built, the floodplain narrowed, and the river straightened trench-like for several miles. Industry soon rose out of the new flood-protected land between the railroad and the new levees. Roadways were then stretched over and under the tracks to reach the new development, some extending as viaducts over the floodway into western Dallas. One of these, the Commerce Street Bridge, gathers three city streets, Elm, Main, and Commerce. As the merging streets intersect the track they dip between two grassy knolls and pass under a three-part concrete railroad trestle known as the "triple underpass." No doubt the structure was in President Kennedy's sights moments before shots rang out of the nearby Book Depository. Another six seconds and Kennedy's motorcade, moving west on Elm Street, would have been under the trestle and out of harm's way. At this hallowed spot, railroad and natural topography fuse as memorial to one the nation's most wrenching events.

Then came the highways. Eighty-seven years after the Golden Spike celebration, President Eisenhower signed into law the Federal-Aid Highway Act of 1956, also known as the National Interstate and Defense Act. It encompasses more than 46,000 miles of roadways, the largest of its kind anywhere. Their mark upon the landscape is grand and indelible. The highways' numbering system gives the landscape both measure and cardinal coherence. Exit numbers relate to miles travelled from a jurisdictionally-based starting point. Sunrises and sunsets can be hastened or slowed, respectively, on even-numbered highways; odd-numbered highways track the commerce between us and our neighbors to the north and south. In north Texas the so-called NAFTA Highway splits into eastern and western legs, becoming 35E and 35W and confounding the sense of orientation of anyone unfamiliar with the Dallas-Fort Worth conurbation.

Plenty of lore is attached to the interstate system by means of a naming sub-layer, a form of infrastructural christening. A section of Interstate 70 in St. Louis, Missouri, skirts the famous arch along the downtown and just past it slings eastward across the Mississippi River. It was named the Mark Twain Highway—quite a leap from his Mississippi rafting days—but only after the steroid scandal that tainted its first namesake, homerun hitter

The American Wilderness: An Evanescing Myth

Mark McGuire, forced a recall. Other highways are named after individuals as different as General Douglas McArthur (I-580 in northern California) and Chicago evangelical minister Bishop Ford (I-94, in northern Illinois). Plenty more invoke the memories of political, sports, and religious figures, activists and law enforcement heroes. Highways, in effect, function as parchment over which the tales of the nation are remembered.

In their scale and extent, the interstate system stands as an outsized monument to the nation's progress—oftentimes dwarfing their surroundings. At interchanges, highways can rise to dizzying heights as multitiered lanes bypass each other with spans that stretch belief in the strength of steel or concrete. In Dallas, the famous "High-Five" stack interchange rises to 120 feet, occupying almost as much land as the city's downtown. Given such impactful grandiosity, it is of little wonder that McHarg would have eagerly consulted with the Bureau of Public Roads during the Lyndon Johnson administration. He was highly contemptuous of highway engineers and their disregard for environment. "Design with nature" was the better way to build the interstate system, a matter that received the president's attention via Lady Bird Johnson[5]. Alas, the advent of mitigating sound walls in the mid-1960s had not yet become commonplace, or else McHarg would have surely railed against them as proof-positive of their hosts' unfitting alignment.

Wilderness Subsumed

I have often wondered where in the United States might be the largest patch of wild land unencumbered by interstate highways. A scan through a map of the interstate system readily draws the eye to the western states. This is "big country" and, compared to the rest of the nation, the interstate highways appear distended, like lines on an expanding air-filled balloon.

The eye then settles on a large polygon between Salt Lake City and Bakersfield, California, bounded by I-80 to the north, I-15 to the south and east, and I-5 to the west. Most of the polygon encompasses the Basin and Range Province, a region described by the US Geological survey as "steep climbs up elongated mountain ranges that alternate with long treks across flat, dry deserts, over and over and over again!"[6] In other words, a repetitive, relentlessly desolate landscape. This is the place that in 1846 claimed the life of one member of the Donner family and left others near dehydration before they ventured up the Sierra Nevada to suffer the infamy of cannibalism. It is the place where the Humboldt River watershed fails to deliver an ocean outlet, the river simply emptying into an interior sink. It is quite possibly the most inhospitable, unaltered landscape in the United States. But perhaps not.

Within the basin and range is Yucca Mountain, a former bombing range and nuclear testing site that has been selected as the repository of the nation's stockpile of nuclear waste. More than 150 million pounds of radioactive material is designated to travel across the nation on rail and over the interstates to be buried deep within the mountain's ignimbrite rock, a form of lithified volcanic ash. Tunnels were dug and the facility prepared to receive the radioactive material. But to-date not one ounce of radioactive waste has been delivered. The Obama administration has suspended funding pending a new evaluation of alternative disposal methods and sites to take the spent nuclear fuel. Such waste may never be stored there. Still, a virtual online satellite tour of Yucca Mountain reveals countless blast craters from a previous era—basins on the range that leave phantom claim to a land far removed from highways and perhaps even farther from being wild.

And so, the boundless and bountiful land in which to roam free in search of opportunity has been subsumed by the national grid, the railroads, and the interstate highways—to say nothing of the national power grid, communications towers, dams, levees, irrigation canals, water tanks, wind farms, and stratospheric contrails that slice through and punctuate the landscape. Few images are more telling than transmission line corridors in snow country. Power companies maintain their easements in clear-cut condition, yielding after snowfall wide, straight, and pure white swaths tens of miles long up the slopes, across contours, and over mountain ridge tops, all in the name of electricity.

Power lines, roads, and rail tracks fan out from urban zones, stretching through suburbia and to the furthest outreaches of development. We draw life from the nation's infrastructure in support of economic production, processing, consumption, and exchange. But we also give it life as we thirst for more blacktop, more tracks, more energy, and more water—more of everything. City and town centers are biomorphic entities, like neurons, big and small, complete with the axons and dendrites of infrastructure. Towns, cities, and infrastructure constitute the nation's brain—the locus and embodiment of a collective intelligence and creative pulse. The gray matter is spread all over, coast-to-coast, from Canada to Mexico, even in sparse North Dakota where a few hundred souls cling to an improbably long and lonely road.

There was indeed a time in the United States when roads and towns were sought-after paths and welcomed oases amid a perceived wilderness. Taming this landscape to forge a nation was a powerful enterprise, requiring equal measures of fearlessness, ruthlessness, perseverance, and optimism. When the spirit of conquest subsided, the landscape assumed a more benign and endearing quality, something to be admired, attended recreationally, even engaged toward spiritual pursuits. But not all landscapes were regarded equally. By virtue of their unique physiography, some were accorded privileged status, such as the Yellowstone caldera and its hydrothermal hot springs. Its preservation as public land was an anthropocentric impulse, a thing to do for personal pleasure and cultural edification.

The first generation of national parks—Acadia, Arches, Big Bend, Crater Lake, and the Great Smokey Mountains—all were established for their unique terrain and beauty. But biocentrism, the ethic dictating that animal and plant life deserved outright protection, regardless of human want, was not far behind. Dedicated by Harry Truman in 1947, Everglades National Park is an emblem of biocentrism. Without topographic drama—no mile-deep canyon, no spectacular waterfall, no volcanic remnant—the Everglades makes its case strictly on the merits of its flora and fauna—for their own sake.

Biocentrism is embedded in the mission of FWS. The Service's editor-in-chief was for many years Rachel Carson. Between 1939 and 1952 she wrote pamphlets and edited scientific articles on natural resources. The epigraph by Rachel Carson at the beginning of this chapter appears on the FWS website; it ends as follows:

> Wild creatures, like men, must have a place to live. As civilization creates cities, builds highways, and drains marshes, it takes away, little by little, the land that is suitable for wildlife. And as their space for living dwindles, the wildlife populations themselves

decline. Refuges resist this trend by saving some areas from encroachment, and by preserving in them, or restoring where necessary, the conditions that wild things need in order to live.[7]

Alas, her call did not slow the progress of "civilization." Through Carson's tenure at the FWS, and beyond to the close of the century, urban land and supporting infrastructure in the United States has more than doubled. Wildlife refuges palliate this condition but, in end, as the John Heinz National Wildlife Refuge exemplifies, these are just safe havens of nature amid a nationwide urban web.

Urban life devoid of access to such safe havens was unthinkable to McHarg. This caused him to travel far and wide to advise governments on how to preserve and showcase their natural heritages. In Iran in the early 1970s he led the planning of Pardisan, an environmental park the size of the John Heinz National Wildlife Refuge, within the Tehran metropolitan area aiming to exhibit the nation's outstanding wild lands. And in Taiwan a decade later he prescribed a process by which that nation could create a system of national parks. The effort directly led to the establishment of Taroko National Park, a preserve half as large as the Everglades National Park highlighted by a deep gorge carved by the Liwu River out of marble formations. These were heroic undertakings: in Iran it cost McHarg parting with his partners over the financial loss stemming from the fall of the Shah in 1979; and in Taiwan it caused the enmity of powerful industrialists and his subsequent dismissal as an environmental advisor to the government of Taiwan. And yet such heroism produced only discrete parches of nature amid widespread development.

It is unquestionable that in the United States, as in much of the world, the pattern of settlement has been forever reversed: the once all-encompassing natural field harboring islands of development has been overtaken by an all-encompassing urban field harboring islands of nature. The nature around us has changed and in turn it has changed us. We are now *Homo Urbis*.

Chapter 3. Cities:
Our Abode

"Are we there yet?"

—Children Everywhere

It takes about eight hours to drive nonstop from Philadelphia to the four-corners town of Greensboro in northern Vermont, then another couple miles to a lone vacation getaway cottage in the midst of rolling farmland. It is my family's tradition to visit this remote destination periodically and enjoy country life far removed from, well, everything. Over 90 percent of the route is driven along interstate highways, the remainder on two-lane county roads and, toward the end, over a winding dirt road past cornfields and dairy farms. The scenic transition over the trip's 440 miles cannot be more stark—from the skyline of the nation's fifth largest metropolis to an elderly farmhouse nestled deep in the bucolic hills of the nation's least populated state. Between these extremes, one encounters every other conceivable development condition, from hamlets to villages to towns to edge cities and lesser metropolises. Most of the route consists of suburbia, however, most unrelenting through the 100-mile Connecticut portion of the journey.

The extraordinary development gradient within this territory speaks of the difficulty federal government agencies have in determining the extent of the nation's urban land. A defining line between what is urban and what is rural no longer seems to exist. The U.S. Department of Agriculture estimates that approximately three percent of the area of the U.S. is "urban," housing 79 percent of the population. The remainder lives in "rural residential areas," comprising four percent of the nation's land area. Rural residential areas are defined as "acres of land and associated lots in rural areas used for housing" or, perhaps more accurately, development occurring in previously farmed land. The rest of the nation's land is classified as non-urban: forests, rangeland, cropland, parkland, military installations, and etcetera.[1]

However, it would be wrong to conclude that 93 percent of the country is somehow free of development. The Office of Management and Budget (OMB) ascribes as "Metropolitan" areas with populations of 50,000 or more, and as "Micropolitan" clusters of development amid farmland or other open land with populations between 10,000 and 50,000. There are 363 "Metro" and 577 "Micro" zones in the United States, the mapping of which suggests at least 50 percent urban coverage within the lower 48 states. Within this coverage, the OMB has identified 125 regions, or Combined Statistical Areas (CSAs) exhibiting strong "employment interchange" (i.e., commuting) ties. At the top of the list is the New York-Newark, Bridgeport, NY-NJ-CT-PA CSA, with a population of about 23 million. At the bottom: the Clovis-Portalis, NM CSA, with a mere 64,000. Almost the entire route from Philadelphia to Greensboro runs through CSA classified areas. Predictably, the journey's scenery is almost entirely developed, the majority as suburbs.

Urban land today defines the American landscape. Much of it exists as a low-density residential mat etched by strip commercial corridors, highlighted, brooch-like, by intermittent hubs of offices and industries. McHarg decried this vast territory not for its scant density, but for its injurious impact upon nature. His mission was to sensitively integrate development with a healthy, natural ecology, preserving valuable natural resources. He maintained that human health (to include a measure of spiritual fulfillment), was to a large extent dependent on ecological health, which he regarded as the greatest source of beauty and the ultimate aim of design upon the land. Like a new and fashionable drug, however, the side effects of suburban living began to be felt decades after its initial intake—especially those related to automobile dependence and associated sedentarism.

Most Americans get to work in an automobile and take to the roads for nearly every other necessity, every day of the week. It is curious that so much of the country's personal mobility depends so utterly on a certified instrument of mass destruction. Every year more than 40,000 people perish in automobile accidents and thousands more are disabled. It's as if several times a week a commercial jet fell out of the sky killing everyone on board. On the rare occasion when this occurs, the news spreads instantaneously, people everywhere become alarmed, images of grief dominate the airwaves, vigils take place on behalf of the dead, and investigations begin in the immediate aftermath. And yet most people regard highway accidents as acts of God, a natural occurrence not unlike the haphazard slaughter that occurs daily among animals in the wild. Even when it happens to someone we know or a loved one, we accept it as a horrible but still random event. Any connection between such a tragedy and the adjudication of land use, development pattern, or the catalytic effect of federal and state funded infrastructure is seldom considered. Most people simply do not see the cause and effect embedded in such policies. Few have calculated whether a single family house with a yard, far from work or mall, is worth the risk of being disabled for life, or worse.

Even so, auto-related casualties impose far less a cost on society than the ill health effects of sitting behind the wheel each day, every week, for hours on end. The Centers for Disease Control and Prevention has established a minimum recommendation for daily moderate-intensity physical exercise. Walking or bicycling to work—even walking a few blocks to a transit station—would advance public health in significant ways. And yet such form of commuting is wholly unattainable to the majority of the population, contributing to the nation's obesity epidemic. In a 2004 study, Franks, Andersen, and Schmid concluded that "each additional hour spent in a car per day is associated with a six percent increase in the likelihood of obesity." As a corollary, "each additional kilometer walked per day is associated with a 4.8 percent reduction in the likelihood of obesity."[2] Obesity is linked to type 2 diabetes and to colon, prostate, and breast cancers. It is also linked to high blood pressure, osteoarthritis, high cholesterol, heart disease, and gall bladder disease. An estimated 300,000 potentially preventable deaths occur each year as a result of obesity.[3]

We will never know how McHarg's approach to regional planning might have evolved had he lived long enough to fully grasp the social and economic impacts of long commutes, traffic fatalities, obesity, and smog-induced pulmonary disease, to say nothing of the effect of vehicular emissions upon climate change. We do know, however, that he regarded the denser realm of cities as unfit human environments. In "The City: Health and Pathology," the last chapter in *Design with Nature*, Philadelphia is examined from an ecological perspective. Included are maps depicting the incidence of various diseases, including tuberculosis, diabetes, syphilis, and bacillary dysentery. To these is added the geography of social ills, such as homicide, alcoholism, robbery, and juvenile delinquency. The "layer cake" synthesis of these conditions coincides with the city's core area, leading McHarg to conclude that density, more than poverty or other social factors, was the cause of such misery. Such analysis at the time amply supported the public perception that urban core areas were unsafe and unhealthy places to live, a quasi-myth every much as powerful as the one that tugged the other way—toward a nature-salved suburbia. It would be nearly a quarter century after the publication of *Design with Nature* that a new model of development—new urbanism—would emerge as a middle ground and, in doing so, at least suggest that living in close proximity to neighbors in denser environments could be a viable way of life.

Densifying the American Dream

A group of architects and planners met at the Ahwahnee Lodge in Yosemite National Park in the fall of 1991 to rethink the prevailing policies and design attitudes guiding the development of urban land. Two years later the group would found the Congress for the New Urbanism, formalizing their quest. It is ironic that their initial encounter would have taken place in the American altar of naturalism (notwithstanding the fact that the land was conceived as pristine after the forcible removal of the valley's native people, the Ahwahneechees, and then only after prospected gold was exhausted). The American wilderness myth pervaded the proceedings, but this time the story was used to promote the creation of a more sensible "non-sub" urban landscape. A manifesto emerged:

> Existing patterns of urban and suburban development seriously impair our quality of life. The symptoms are: more congestion and air pollution resulting from our increased dependence on automobiles, the loss of precious open space, the need for costly improvements to roads and public services, the inequitable distribution

of economic resources, and the loss of a sense of community. By drawing upon the best from the past and present, we can, first, in-fill existing communities and, second, plan new communities that will more successfully serve the needs of those who live and work within them.[4]

Seven community design and 15 regional planning principles were crafted in support of the manifesto, becoming virtually synonymous with Smart Growth. Three are worth noting:

- All planning should be in the form of complete and integrated communities containing housing, shops, work places, schools, parks, and civic facilities essential to the daily life of the residents.
- As many activities as possible should be located within easy walking distance of transit stops.
- Each community or cluster of communities should have a well-defined edge, such as agricultural greenbelts or wildlife corridors, permanently protected from development.[5]

The first principle promotes walkability by means of a mix of land uses within proximity of each other. Walkability has a direct positive impact on obesity. The second addresses the environmental and health benefits of commuting to work by means other than the automobile. Using public transportation to get to work reduces carbon emissions, relieving a host of respiratory ailments. And the third places emphasis on open space conservation, a key variable in determining the form and intensity of urban development. The latter is clearly in line with McHarg's environmental ethos. It was in fact the blending of his ethos and that of the Ahwahnee conferees, namely environmentalism and (new) urbanism that was operative in the formulation of the South Livermore Specific Plan mentioned in Chapter 1. The result, as described below, is representative of the unavoidable shift toward a more compact form of development, driven by the desire to conserve open land as part of a sustainable, smart growth agenda.[6]

Lying 30 miles east of Oakland, the South Livermore Valley is famously known for its vineyards, which rise and fall over gently rolling hills like

Reconsidering Ian McHarg

Figure 3.1. New compact development in seven distinct parcels adjoin the town of Livermore, California. Outlying open space and agricultural lands are preserved in perpetuity through a transfer of development rights.

waves of calico green. It is anchored by the town of Livermore, home to the Lawrence Livermore National Laboratory. For decades, development had crept languidly along I-580, but in the early 1990s, development reached the doorstep of the valley's finest vineyards in the form of Ruby Hill, a gated community offering 680 luxury homes anchored by a Jack Nicklaus-designed golf course.

Fearing the spread of similar development deeper into the valley, 22 landowners joined forces in support of a development plan that could preserve the area's unique rural heritage. That WRT was retained for the assignment was largely based on the belief that "design with nature" would deliver both the method and ethos by which the valley could be simultaneously developed and protected. But also present were expectations for a new urbanist approach favoring tight-knit neighborhoods in close proximity to services.

To meet these dual objectives, the Specific Plan allocated approximately 1,200 by-right single-family dwelling units into seven discrete parcels representing a quarter of the project area—a fourfold densification, made possible through a transfer of development rights mechanism. Within each of

CITIES: OUR ABODE

the seven parcels, a deliberate arrangement of traffic-calmed streets, alleys, sidewalks, trails, and greenways facilitate walkability and cycling. The largest of them adjoin existing urbanized land (see figure 3.1). In essence, the plan expanded the existing townscape rather than growing its boundary into the hinterland in haphazard fashion. Overall, 850 acres of hillside agricultural land was preserved open and productive, much of it contiguous and in perpetuity. Working vineyards and olive orchards were introduced within the fabric of the development to stamp the rural character as a defining quality of the development (see figure 3.2). Additionally, in collaboration with architect William Turnbull, the northern California farmstead vernacular was encoded in the plan as the guiding residential style. (Bill Turnbull was a long-time admirer of McHarg. In collaboration with Lawrence Halprin and former partner Charles Moore, Turnbull had designed Sea Ranch, a community noted for its "fitness" with respect to Sonoma County's harsh, windswept coastal environment. The development's decumbent and shingled homes exemplified Turnbull's adaptation of the vernacular as a source of truth in architecture. To

Figure 3.2. In Livermore, new homes are nestled between preserved hillsides and a productive landscape of olive groves and vineyards, which serve also as buffer from roadways and recreational trails.

him, the vernacular represented irrefutable fitness between man and nature, a concept bordering on the sacred that could well have leapt out of the pages of *Design with Nature*. His work on the *Specific Plan for the South Livermore Valley* was to be his last major achievement before succumbing to cancer in June, 1997.)

The South Livermore Specific Plan represents but a minute example of densification in exchange for open space. Nationally, the trend to conserve open space is substantial and growing. The Trust for Public Land reports that since 2005, approximately 10 million acres have been protected from potential development through conservation trusts and easements involving federal, state, and local initiatives (23 million since 2000).[7] Of these, 375,000 acres are in the state of Texas, 530,000 in Arizona, 170,000 in Florida, 246,000 in Georgia, and more than two million in California. These are Sun Belt states that in the past decades have attracted a large share of the nation's urban growth. And this does not include development-instigated land protection such as that in the South Livermore Specific Plan, where environmental as well as scenic concerns led to a community-led conservation of 44 percent of the developable land.

The Northeast also has robust open space conservation programs. In Baltimore County, for example, easements established under The Rural Conservation Program protect more the 50,000 acres from potential development, widely expanding the conservation of open land initially conceived in *The Plan for the Valleys*.[8] The County's urban-rural demarcation line exists in support of such rural conservation programs, helping also to contain growth into a tighter ring around the city of Baltimore itself.

In line with smart growth principles, the drive to conserve open land (however crisscrossed by infrastructure it may be!) will effectively curb the extension of urban development. Los Angeles, Phoenix, Dallas, Atlanta, and others will not duplicate in area as their populations grow. Arguably there will come a time when urban land will be hemmed in by open space conservation easements; when, as a result, leapfrog patterns of development will no longer be possible; when urban regions will be contained by broad agricultural and wilderness greenbelts; and when, given such limitations, new development will have nowhere to go but upward rather than outward. One needs to look no further than South Florida to envision such a future.

CITIES: OUR ABODE

UP, UP, AND (NOT) AWAY

Southeast Florida is hemmed in between the Atlantic Ocean on one side and the Everglades on the other. The region is virtually built up. Barring the filling of Biscayne Bay or convincing the federal government and the South Florida Water Management District to move the urban growth boundary into the Everglades, the Miami region can only continue to develop vertically.

The trend has been underway for several decades. It was ushered in during the early 1980s by Park West, a new-town/in-town, a planned 86-acre mixed use project occupying marginal industrial lands just north of downtown Miami (see figure 3.3). Three thousand dwelling units in high-rise apartments underlayed by structured parking and ground level retail were drawn up in the hope of bringing new residents and businesses into the city's core area. A half-mile long promenade with shade trees and fountains offered a common amenity, linking Biscayne Bay at one end of the project with a Metrorail station on the other. (The plan was led by WRT under the direction of David Wallace. It was commissioned by the Downtown Miami Development Corporation.) Shifting political

Figure 3.3. As envisioned in the early 1980s, Park West was planned for 7,000 residents within a few blocks of downtown Miami.

39

Figure 3.4. High-end condominiums facing Biscayne Bay (background) and affordable housing facing historic Overtown rise as bookends in Park West today. The middle ground awaits redevelopment.

and economic winds ultimately modified and deferred much of the plan's implementation. But vertical densification did nonetheless occur: a block of affordable high-rise housing near the Metrorail station facing a rambla-styled promenade, and several blocks of towering condominiums facing the Bay, which significantly extended the downtown skyline northward (see figure 3.4).

Other nodes of development intensification have followed suit. Dadeland, initially a free-standing suburban mall surrounded by a sea of parking nine miles south of downtown, is today a thriving, high-density town center with offices, high-rise apartments, shopping and entertainment venues linked to the Metrorail. Surrounding the town center, mid-rise apartments and townhomes have multiplied. And mid-point between Dadeland and downtown Miami is Coral Gables, a once placid residential community that today is anchored by a downtown of its own. Mid-rise buildings stretch for several blocks north and south of Miracle Mile, the town's historic Spanish-style main shopping street. Their bulk stands in sharp contrast to the town's low-scale grain of residential development, posing a regionally-scaled visual landmark. Since 2005, a 50-acre, high-density new community began emerging two miles north of Park West over a former railyard. Called Midtown Miami, the mixed use development will eventually have eight residential towers. Shops, restaurants, entertainment and cultural attractions occupy lower building floors, giving the development day-night vitality (see figure 3.5).

As a result of development intensification, South Florida today exhibits a fairly high population density. At 1,230 people per square mile, the Miami-Fort Lauderdale conurbation is almost as dense as Chicago's (1,322) and denser than that of Boston (1,034). It is also significantly denser than other major sunbelt cities, such as Atlanta (671), Dallas-Fort Worth (573), Phoenix (223), and Riverside-San Bernardino (119). Not surprisingly, 57 percent of Miami's population resides in multifamily housing while in Riverside, California, only 30 percent do.[9] This is logical: without defining natural or man-made conservation barriers, these other regions have enjoyed virtually unbound land for single-family urban growth.

Figure 3.5. Midtown Miami, a mixed use development rising over a former railyard two miles north of downtown, embodies Miami's emerging higher density urban lifestyle. A 3.5-acre park is planned as a central amenity.

But these sprawling cities, too, are finding their limits and are beginning to add growth from within. In Atlanta a grassroots effort has led to the repurposing of a 20-mile rail corridor encircling the city's core area as a redevelopment catalyst. The "Beltline" is envisioned as a green corridor linking multiple walkable and transit-oriented mixed use developments. In North Texas, both Dallas and Fort Worth are transforming large areas of the Trinity River Corridor into major urban parks to effect the redevelopment of marginal industrial lands close to their respective downtowns (see Chapter 8). Tempe, Arizona, constructed a pneumatic damn on the Rio Salado in the late 1990s, causing a lake to rise as a central amenity for future high-intensity development. Several mid-rise projects have gone up facing the artificial lake. And in Houston, the "Buffalo Bayou," a ten-square-mile area bordering a major floodway and adjacent to the downtown, is planned to become a mixed use district with tens of thousands of residents. These are all mid- to high-density development initiatives with light rail, rapid bus, or streetcar programs in the mix.

But if we build it, will they come—*in the numbers necessary* to make a defining difference over sprawl? Most Americans still regard dense urban areas as undesirable places to live and will work there only if they can readily retreat back to their suburban lairs. If the U.S. Congress suddenly mandated urban densification, requiring families to live in a townhome or apartment, there would be massive political upheaval. And for good reason: most urban areas remain stubbornly noisy, dusty, grimy, polluted, and lacking in greenery. But what if cities had more trees and gardens, their water and air were clean, and urban noise was dampened. What if there were safe and easy access to greenways, parks, and play areas for children close to every home, green rooftops for those living in high rises, streets that were cool and shady with wide sidewalks inviting people to stroll, meet, and chat, bicycle lanes to every destination, and the built landscape suffused with art and craft—the kind of design and materiality that people would notice and care about? What if all of the above occurred in denser, transit-oriented regions that rendered automobiles not a necessity but rather an occasional convenience?

Perhaps, toward the beginning of the next century, urban historians will look back and pronounce the passage from the second to the third millennium as a time when Americans saw fit to create their urban domain anew, impelled by the necessity to house more of the population in fewer acres. McHarg's vision of a protected nature in which to sensitively integrate development is hardly applicable to this agenda. Yet we should still heed his call for environmental quality, especially so in the context of densifying core urban areas. As argued herein, such quality should accrue, among other factors, from the integration of a working nature (green infrastructure), community rootedness and engagement (localism), and meaningful craft (public art). Ultimately, the coalescence of the city-making professions, from planners and urban designers to architects, landscape architects, civil engineers, and artists, must create a distinct urban milieu—*a new urban nature*—that proves superior to the idyllic setting of suburbia. A sustainable future demands as much.

Chapter 4. Building, Dwelling, Greening

"The possible ranks higher than the actual."
　　　　　—Martin Heidegger, *Being and Time*, 1927

Mobius Strip: "a one-sided, non-orientable surface."

Imagine yourself during a summer outing stopping at vista points along the Blue Ridge Parkway. The sky is clear and the mountain landscape is broad, stretching far into the distance. The near slopes are green in hue and coarsely textured by a canopy of hickory, tulip poplar, chestnut oak, and sycamore. As the ridge tops recede, the pointillist foliage turns an impressionistic sky-laden shade of blue. You lace up your hiking boots and head for the trailhead. Soon you are immersed in a world of dappled shade, myriad leaves yawing lazily above. Tree limbs slice through the foliage catching fragments of light that pierce what moments earlier seemed like an impenetrable canopy. You sense the murmur of a stream. You walk further and encounter crystalline water weaving and tumbling gently over moss-covered boulders. You sit to rest and absorb the never-ending dance between the elements: water, the ultimate transmutable substance, and rocks—rounded feldspar—the very definition of intransience. Deep in the woods you are struck by the realization that you have not left the surface of things, only transitioned seamlessly from the outer profile of rolling mountains to the inner complexion of leaves, twigs, trunks, boulders, pebbles, and creeks that define Appalachia.

And then you ask: what system supports such wonder? How is it organized? You have a basic understanding of ecology: it is the science of nature, of the relationship between living things and their environment. You are reminded that in an ecological system every part, however small, plays a role in support of the system's overall vitality: that green chlorophyll pigments in leaves absorb sunlight and use it to synthesize carbohydrates from CO_2 and water; that leaves fall and over time decompose, attracting nitrogen-fixing bacteria that enrich the soil; that soil, granular and malleable, is a sponge-like substance that holds water; that water, through turgor pressure, is the medium by which tree roots absorb the nitrogen that feeds the foliage; and that foliage, through evapotranspiration, returns moisture to the air and transforms Appalachia into a carbon sequestering, water recycling, and solar-collecting field extending in all directions as far as the eye can see. This is a working nature that never sleeps, processing CO_2 into oxygen in a state of constant regeneration, nurturing creatures large and small, from the black bear to the brook trout—*and you, homo urbis, curator and guardian of it all* (see figure 4.1).

Breathing Green

Back home, early in the morning, you walk up to the building rooftop to greet the emerging cityscape. From the compound's greenhouse you

Figure 4.1. In a conservation area in Laporte, Pennsylvania, rock, water, fungi, moss, bark, twigs, leaves, and wildlife comprise the finer grain of Appalachia's never ceasing ecological service.

pick a few tomatoes and some basil. The productive roofscape is irrigated by recycled water, biologically-purified by an array of "living machines." Throughout the complex, every drop of water is captured, naturally treated, and recycled without leaving the premises, dispensing with the collection and conveyance of sewage to a distant treatment plant. The forecast calls for warm temperatures but your home will remain comfortable as always, moderated by a north-facing water wall that scoops outside air and spreads it like a cooling front throughout your living quarters. Broad south-facing windows bring natural light into the family room, shielding it from direct sunlight by an overhead grapevine that traps airborne dust while reducing bothersome glare. You scan neighboring terraces for signs of life. Each one is lushly planted with leafy matter which, from a distance, fuses with the vegetated walls that cover much of the architectural walls and interstices. The scenery reminds you of Appalachia—fold upon fold of living textures that, like a Mobius strip, remain surface-bound.

After breakfast you walk downstairs with the rest of your family to retrieve the bikes. As you exit the building you notice the dew on the vegetated wall, droplets glistening in sunlight. The wall system is eight inches thick, enough for the vegetable mass to provide superior insulation and sound attenuation. Rooted in coir-packed crates, the plants also function as a carbon sink, ten square feet of surface area, enough to sequester as much carbon as a medium size tree. You recall a time when green walls were an architectural novelty, but no more; they are now as ubiquitous as brick facades once were and every bit as essential for public safety—erstwhile bricks to prevent the spread of fire, green walls to do as much plus conserve energy and scrub the air.

Heading uptown, your spouse sets off on the cycle track paralleling the transit-way. You and the girls jump onto the greenway trail for a quarter mile or so. An overnight downpour has filled the stormwater retention areas at either side of the path. The wetland is the terminus to the network of rain-gardens, bioswales, channels, and catchments that filter storm runoff before any overflow reaches the dips and hollows that comprise the greenway landscape. As you approach the crossing creek you signal your crew to stop. You've spotted a bald eagle perched on a nearby branch at the far bank of the watercourse. You look up as other cyclists catch your gaze and direct their eyes skyward. The bird remains unperturbed, seemingly more interested in the goings-on in the water below than in any human commotion. You remind the girls that not long ago the creek was an encased sewer, buried and paved over, but that under the state's 2040 Vision Plan it was dug up and daylighted, the 150-year old brick-lined pipe crushed and recycled as a growing medium for the community's hydroponic farm co-op.

As you near the neighborhood elementary, you recall how hard it rained the night before and tease the kids by insisting that the schoolyard will surely be flooded. They chuckle. Paved with porous asphalt, the yard is bone dry. As always, children are running around, playing tag, bouncing balls, waiting for the bell to ring. A ball strikes a solar-fabric-covered light pole, but the rugged material easily absorbs the impact. Many of the outdoor surfaces throughout the learning campus produce electricity, adding to the photovoltaic panels that blanket the buildings' vegetated rooftops. Each classroom sports a voltage meter showing how much energy at any given time the school sends to the power grid; it rarely goes negative, but when it does, the lights are dimmed, computers are turned off, and sessions of storytelling begin.

You hug the girls goodbye and pedal away, riding over the boardwalk edging the school's biotreatment ponds. The sedges and lilies in the terraced ponds absorb phosphates, nitrates, and residual coliform bacteria, yielding the water that irrigates the playfields. Solids are separated prior to the flow of effluent into the ponds and are heat-composted along with leftover cafeteria organic waste for use as field and garden fertilizer. You glance toward the fields and marvel at their verdant sheen. Soon you are coasting down a sharrow, a marked lane shared by both cyclists and motorists. To your right are tufts of native grasses rising out of the curbside rain gardens. The street is shaded by broad deciduous trees that keep much of the sun off the roadway, mitigating heat absorption and glare. You feel the breeze flow through you as you glide past the last block to your office building. You park the bike and pause for a moment at the entry court. You are pleased to live in a green world.

The preceding is an imagined world, one that McHarg hardly envisioned. Even today, no urban community is fully powered by renewable energy, affording naturally scrubbed air, recycled water, bioregulated drainage, landscape-moderated microclimates, natural carbon sequestration, locally produced foods, bicycle trails to every destination, and abounding wildlife in its midst. But it is an eminently possible world, one that would render America's inevitable urban densification healthy and desirable. Each green technology mentioned above is in existence today and proven to work, from the localized treatment of sewage through biotrophic means to multilevel vegetated walls and solar fabric-wrapped utility poles. What has not yet occurred but can easily be imagined is the comprehensive application of green infrastructure at the metropolitan scale—much as the interstate highway system existed as a latent plan until Eisenhower pushed for its comprehensive spread over the landscape.

From Design with Nature to the Granite Garden

It is fitting that major advances in the conception and application of green infrastructure at the metropolitan scale have occurred in dense, brownfield Philadelphia, seat of the program in ecological planning and design that McHarg guided for years to the benefit of greenfield development well outside its urban boundaries.

McHarg did not ignore his home base; he studied it, rather, to prove the cause-and-effect relationship between density and environmental and social degradation. The remedy to such squalor, he asserted, was the need to know "where the environments of health are, for there the environment is fit and the adaptations are creative."[1] By this he meant the nature-abounding region outside the city. Seemingly precluded was the possibility that human intervention could repair unhealthy urban environments and make them fit and desirable places to live and work. Scattered in other writings McHarg approaches the subject of green infrastructure. In an essay entitled "The Place of Nature in the City of Man," he points to the use of water resources, floodplains, wetland, woodland," and even farmland as part of a "matrix of natural lands *performing* work or offering protection and recreational opportunity distributed throughout the metropolis" (author's italics).[2] But McHarg's reference to what today is understood as ecological service is restricted to large-scale natural features; little is offered about possible methods by which nature can perform work at the micro-scale of building rooftops, gardens, streets, or even neighborhood parks.

But where McHarg fell short, one of his esteemed students, collaborator and eventual successor as chair of the department, Anne Whiston Spirn, did not. (Spirn became chair in 1986, a position she held for eight years.) Her academic focus became the study of urban ecology as a way to address environmental and social degradation. In 1984 she published *The Granite Garden: Urban Nature and Human Design*,[4] a seminal work on the subject. The book opens with an impassioned salvo:

> Nature pervades the City, forging bonds between the city and the air, earth, water, and living organism within and around it. In themselves, the forces of nature are neither benign nor hostile to humankind. Acknowledged and harnessed, they represent a powerful resource for shaping a beneficial urban habitat; ignored or subverted, they magnify problems that have plagued cities for centuries, such as floods and landslides, poisoned air and water. Unfortunately, cities have mostly neglected and rarely exploited the natural forces within them.[4]

Issues of air and water quality, heat island effect, storm drainage, flooding, water supply, energy efficiency, and urban vegetation and wildlife are addressed in the book, laying a broad foundation for rethinking the relationship between nature and urban land. West Philadelphia served as Spirn's laboratory, specifically the area's Mill Creek Watershed. Lying across the Schuylkill River from Center City, West Philadelphia is among the nation's early streetcar (initially horse-drawn) suburbs. Large Victorian townhomes with ample street-side gardens were first built, a far closer approximation to Penn's vision of a "Green Country Towne" than at the time existed in the city's older core between the Schuylkill and Delaware Rivers. In later years smaller worker townhomes appeared further to the north and west. Mill Creek courses through both realms, emptying into the Schuylkill River near historic Woodland Cemetery.

First through Penn and later MIT, Spirn for over two decades has led successive waves of students in documenting and explaining the history of Mill Creek and its impact on the local community: how in the eighteenth century colonial industry tapped its water power; how, as the city grew westward in the nineteenth century, the floodplain was filled and the creek compressed and buried into a 20-foot diameter pipe; how in the twentieth century a mixed race community suffered white flight and, in its wake, left behind a social vacuum filled by low-income African Americans and an impoverished urban landscape; how, unaware of unstable conditions, high-rise public housing was built over the filled floodplain only to fail structurally and be demolished.

(Designed by Louis Kahn in 1954—two years after McHarg arrived at Penn—the Mill Creek Housing Project was hailed by critics as a modernist icon very much worth saving. McHarg admired Kahn and his work, qualifying his search for formal purity as prequel to the ecological method—a way to accord inert materials with the "will to be" in the same fashion as the ecological method reveals causality: "*the place is because*."[5] Curiously, just as Kahn's inglorious public housing in West Philadelphia is seldom included as part of his legacy, McHarg appears to have ignored its most un-ecological placement in extending praise to his Penn colleague[6]); and then—stunningly—new low-rise public housing was built in its place with roadways designed to meet modern green infrastructure methods, only to have the city's streets department opt instead for a "true and tried" grey stormwater system. This last chapter in the history of Mill Creek is noteworthy, for had the housing project been delayed a few years, new political leadership would have pressed for green stormwater solutions.[7]

Students from a local middle school have participated in the study through class projects, gaining valuable insight into the value of acquiring landscape literacy. Spirn reports:

> Teachers would take kids and bus them out to Andorra Nature Center. What they were seeing there were many of the same plants growing outside the school. It's nature when they see it out there, but not nature when they see it a block from the school. To me it was tragic.[8] (See figure 4.2.)

Children in the Spruce Hill neighborhood of West Philadelphia attend the Saddie Mosley Tanner Alexander Elementary School, a new public facility that abuts the buried Mill Creek along 43rd Street. Two recent graduates have

Figure 4.2. The Aspen Farms Communty Garden lies over a portion of buried Mill Creek. In 1988–89 it became a focal point of Anne Spirn's West Philadelphia Project, redesigned and improved with the input from neighborhood elementary and high school students.

as a parent Howard Neukrug, current Philadelphia Water Commissioner and former head of the city's division of watersheds. Neukrug is well acquainted with Spirn and the West Philadelphia Project. Inspired by her work and with coincident aims, Neukrug has been instrumental in steering the city toward the use of sustainable infrastructure, especially in new public schools.

This new school supports an interpretive rain garden. Lying almost directly over the sewer, it receives the overflow of the playfield's gravelly underlayment, which in turn functions as a reservoir for the school's stormwater runoff. Adjacent to the rain garden is the "blacktop," a small patch of porous asphalt that serves as play and student drop-off area. In combination, the blacktop and rain garden are evidence of Philadelphia's new outlook as a green city (see figure 4.3). From time to time children and parents amble towards the school's rain garden to picnic, or simply to soak in the lushness and for a while imagine: what if Mill Creek had not been tunneled and were instead a greenway to the Schuylkill River? What if schoolchildren had as their backyard a real urban ecology to study rather than a demonstration garden? What if the city of Philadelphia had embraced the Granite Garden when it was first published? How much greener and more sustainable would it be today? What if the nation as a whole were to mobilize and make urban environments healthy because it is cities, not the countryside, "where the environment is fit and the adaptations creative?" as McHarg noted.

GreenPlan Philadelphia

The "new political leadership" mentioned above refers to Mayor Michael Nutter and his administration. Intending to make Philadelphia "the greenest city in America," the mayor established the Office of Sustainability soon after assuming office in 2008, thereafter endorsing Greenworks as the blueprint for implementation of the city's sustainability agenda.[9] The foundation for Greenworks and other sustainability initiatives was GreenPlan Philadelphia, a comprehensive vision for the city's open areas—both public and private—with green infrastructure at the crosshairs.

The genesis of GreenPlan is credited to the Open Space Planning Group, a joint task force representing city departments and nongovernmental organizations. It was established under former Mayor John Street to address the protection and potential expansion of community parks and open spaces. That initial effort led to the funding in 2005 of an open space

Figure 4.3. Children gather at the pervious schoolyard before classes begin. A rain garden frames the space beyond, the only reminder of Mill Creek's once lush course to the Skuylkill River, three-quarters of a mile away.

pilot study, and a year later the formulation of GreenPlan itself. (GreenPlan was commissioned by the city of Philadelphia Water Department, Division of Watersheds. It was prepared by WRT as lead consultant, under the direction of principal Mami Hara.)

Since the late 1990s Philadelphia has been adding population—30,000 in the past ten years according to the 2010 Census—a modest but welcome reversal of a half-century decline that saw more than a half million residents leave the city. The growth of employers such as the University of Pennsylvania and Temple and Drexel universities, as well as prominent health care institutions, has helped catalyze the resurgence of neighborhoods on both sides of the Schuylkill River and well into the northern areas of the city, a prospect barely imagined in the 1970s and 1980s. GreenPlan Philadelphia supports the revitalizing trend by envisioning a unified, green approach to the city's urban space:

> Rivers lined with new development and necklaced with greenways, trails, and natural preserves, places that nurture wildlife and restore fish as they clean the air we breathe and the water we drink; and tree-lined streets, planted sidewalks, and multi-modal boulevards where bikes, cars and all forms of transportation combine to save energy, increase property values and safely move Philadelphians from place to beautiful place; and parks, large and small that serve to unite communities, foster pride and ensure safety while providing the means for active, healthy lifestyles.[10]

Such a broad-based review of the function and value of Philadelphia's open space had not occurred since the mid-nineteenth century when Fairmount Park was established to help protect the Schuylkill River as the source of the city's drinking water. GreenPlan surpasses that pioneering effort by identifying and quantifying the function and value of all of the city's open spaces—parks of course, but also schoolyards, vacant lots, waterfronts, even currently paved plazas, streets, and transportation corridors (see figure 4.4). Its innovation lies in the integration of objectified ecological service—a "working nature"—into the matrix of urban spaces, along with the identification of environmental, economic, and quality of life ben-

Building, Dwelling, Greening

Figure 4.4. GreenPlan Philadelphia, as depicted in this enhanced photo, addresses the city's public and private space as a comprehensive green infrastructure system; it serves as a foundation of Greenworks, Philadelphia's sustainability blueprint.

efits. This three-part model, summarized in figure 4.5, reveals the interrelated nature of the considered elements.

The filled circles show the plausibility of an environmental, economic, or quality-of-life benefit associated with a green infrastructure element or type of urban space. These are promoted as reasonably achievable targets in comparison to baseline measures (as shown in figure 4.6).

These benefit targets are goals to aim for, to highlight in pursuit of funding sources, and to brandish in the interest of promoting the city as a desirable place to live and work. But they also point to potentially very real, positive, and measurable impacts to the city's overall environmental health.

Carbon Sequestration

With a population of about 1.5 million, a doubling of the number of trees per inhabitant, as called for in GreenPlan, equates to the addition of 2.25

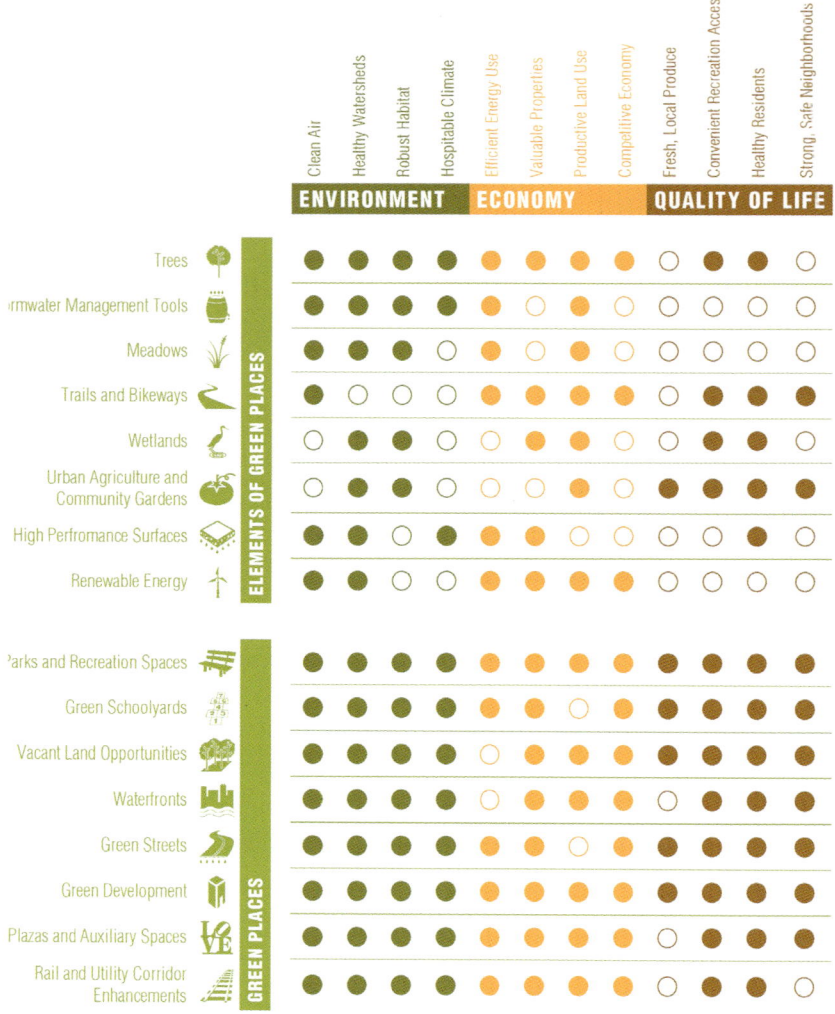

Figure 4.5. GreenPlan Benefits Table.[11]

million trees. The growth of such biomass would sequester about 22,000 tons of carbon dioxide in any given year (using 20 pounds of CO_2/year as a tree's expected absorption rate). Based on the EPA's standards for average vehicle emissions (19.4 pounds of CO_2 per gallon), miles per gallon (20.3), and annual passenger car miles traveled (12,000), such a reduction in carbon equals the impoundment of 3,800 automobiles or, conversely, the

Benefit	Baseline	Target
Clean Air (days)	354	365
Percent of space managing first inch of rain	48	63
Acres of green roofs	4	67
Miles of green streets[1]	0.06	1,337
Acres of forest	6,746	7,200
Acres of wetland	1,086	1,286
Trees per resident	1.5	3
Percent tree canopy cover	13.82	30
Number of urban agriculture businesses	14	24
Percent of residents within a 10 minute walk of a park	58	100
Percent of people within a half-mile of a major trail or greenway	12	100
Miles of bicycle lanes	196	300
Acres of parks per 1,000 residents	7.1	10

1. Streets capable of retaining and filtrating the first inch of rain by means of porous paving, rain gardens, or bioswales.
2. Estimated existing average city-wide tree cover.

Figure 4.6. Objectified ecological service

addition of that many automobiles onto the city streets without exacerbating emission impacts.[12] The global impact of such localized sequestration is very small, yet it points to the potential cumulative impact tree planting programs could have if spread across the nation's cities.

But where could more than two million more trees be planted? Philadelphia does not have a tract of open land large enough. Trees would have to be introduced surgically into the city fabric—on patches of open land wherever available such as small neighborhood parks, schoolyards, sidewalks, street medians, drainage swales, parking lots, utility easements, and vacant lots. To this end, South Philadelphia provides a virtual blank canvas. The greenest blocks in this tightly rowhouse-packed community have three or four trees at best. Most have none at all, and there are few parks within which to expand the tree cover. A street tree program, therefore, would have to be largely implemented within the area's 200 miles of right-of-way. Using 20-foot average spacing as a guide, South Philadelphia sidewalks could support the introduction of about 100,000 trees. Even if only half as many were planted in consideration of obstacles such as underground utilities, overhead power lines, driveways, and other constraints, 50,000 trees would still constitute a significant heat island-mitigating, climate moderating, habitat creating, and groundwater absorbing working nature. All the more so considering the canvas: a mostly barren landscape that soaks and radiates summer heat while subjecting basements to periodic flooding due to high groundwater elevations.

As part of DesignPhiladelphia 2011, a 10-day city-wide showcase of design flair and innovation, SOSNA, a South Philadelphia neighborhood association, constructed a temporary demonstration of street greening and traffic calming measures. Street trees and planted intersection bump-outs

Figure 4.7. During DesignPhiladelphia week 2011, several South Philadelphia blocks were fitted with temporary trees and traffic calming planters as a green infrastructure demonstration. The effort was sponsored by the South of South Neighborhood Association (SOSNA).

were among the demonstration features (see figure 4.7). It was a resounding success, one that received local as well as national recognition. A blueprint was thus established for the application of green infrastructure in a dense urban area. If a tree-planting program can work in South Philadelphia, it can work any place where every square foot of public space is constrained and contested.

Stormwater Management
Philadelphia streets and sidewalks account for 27.5 percent of the city's urban area, or about 37 square miles. A one-inch rainfall over this area represents approximately 2,000 acre-feet of water, or nearly 660 million gallons. Many more million gallons accumulate on hard surface areas in private lands, rooftops, and parking lots. Approximately half the city is served by combined sewers—18.3 square miles producing 975 acre-feet of first-inch stormwater. Collected during rain events through gutters, drainage inlets, and curbside grates, much of this volume flows unimpeded into the Delaware and Schuylkill Rivers through combined storm sewers, along the way gathering raw sewage, oils, chemicals, and other toxic wastes that homes and businesses habitually flush down their drains.

In an effort to stem the impact upon the region's waterways, the city of Philadelphia and the Pennsylvania Department of Environmental Protection (DEP) have agreed to rely, to a great extent, on green infrastructure as the means to retain and filter up to 85 percent of this volume; that is, by diverting the first inch of storm runoff onto green roofs and into rain barrels, rain gardens, bioswales, wetlands, recreational fields, and other pervious surfaces such as porous asphalt, porous concrete, and reticulated pavers. Pursuant to this goal, the city has initiated the repaving of many municipal game courts with porous asphalt. It has also embarked on a Green Streets program designed to introduce green infrastructure measures such as rain gardens and bioswales into municipal rights-of-way.

The resulting Consent Order & Agreement with the Pennsylvania DEP and the Administrative Order for Compliance with the EPA places Philadelphia at the forefront of the green infrastructure movement. As an implementation tool, the water department has instituted a new rate structure for water services, one based on the generation of stormwater runoff

Figure 4.8. Green roofs over Comcast Center utility buildings on Market and 18th Streets provide needed street level vegetation while advancing the city's green infrastructure agenda.

as opposed to the sole metering of potable water. The proverbial carrot is thus dangling in front of private, public, and institutional landowners: treat stormwater flows on site, ideally through green infrastructure means, or suffer higher water bills.

To the area's big box retailers or auto dealerships—businesses that consume relatively little water but generate vast amounts of storm runoff—Philadelphia's new water policy poses a significant incentive to find green solutions to their vast acreage of roofs and parking lots. By contrast, developments that consume larger amounts of water but generate comparatively little storm runoff, such as Center City skyscrapers, will likely see a reduction in their water bills. Water conservation was a major objective in the design of LEED-Certified Comcast Center, the city's newest and most iconic skyscraper. Among many green measures, the building boasts hundreds of waterless urinals. But the project's most readily perceived commitment to environmental stewardship is about a third acre devoted to green roofs over adjoining lower level service areas. This greenery is in plain view of adjacent office buildings, a portion even from street level. Designed by the Olin Partnership, this landscape is a gift to the eye, a green and lush horizontal façade that easily defeats the otherwise dull surroundings of gray paving, steel, and glass. It's as if a living sponge had found a home amid the city's impervious crust, becoming aglow through its cleansing act (see figure 4.8).

Watershed Restoration
Historically, the Philadelphia landscape was riddled with waterways, a pattern of dendrites feeding five major creeks: the Wissahickon, emptying into the Schuylkill River and the Darby-Cobbs, Pennypack, Poquessing, and Toocany-Tacony/Frankford, emptying into the Delaware River. The nineteenth century zeal to remove the visible flow of wastewater from urbanizing areas caused the burial of much of the city's drainage network. The Mill Creek sewer is a case in point. Through its vision of *"rivers that nurture wildlife and restore fish as they clean the air we breathe and the water we drink"* (my italics), GreenPlan brought focus to the restoration of the city's urban waterways, a shift from decades of waterway neglect and abandonment.

Embedded in the water quality imperative is also the desire to reconstitute natural ecologies for recreational purposes. The restoration of a 900-foot section of the Wissahickon Creek in the upper reaches of the watershed in adjoining Montgomery County serves as an example. Within Philadelphia's Fairmount Park, the Wissahickon Creek affords a virtual immersion into a steeply carved beech-maple forest, a remnant of Pennsylvania's sylvan past. But upstream of Fairmount Park the creek is subsumed by a suburban landscape, coursing past golf courses and shopping malls until reaching the site in question where for decades hardly a trace of a natural ecology could be detected. The restoration effort diverts 1.25 acre-feet of storm runoff into a series of bioswales and wetland basins, naturally retaining and filtering water that would otherwise flow untreated into the Schuylkill River 22 miles downstream. (The project was designed by WRT in collaboration with Metz Engineers and Conestoga-Rovers Associates; it was funded by the Pennsylvania Department of Conservation and Natural Resources and the USEPA, along with contributions from the borough of Lansdale and County and Tree Vitalize, a nonprofit organization aiming to add one million new trees throughout the Philadelphia metro area.) In the context of Philadelphia's metro area, the project's environmental and recreational contribution of 14 new trees, 8,000 square feet of wetland, and 2,000 feet of trails, plus a few acres of riparian meadow, is molecular in scale. But it turned a small patch of nowhere suburban land into a productive ecology with demonstrative recreational value.

In Philadelphia, the Office of Watersheds has led the restoration of Cobbs Creek, the city's longest tributary and principal environmental and recreational resource for western Philadelphia neighborhoods. Among the restoration measures are newly stabilized banks, riparian vegetation, and improved access to a trail and play areas (see figure 4.9). Much of Cobbs Creek courses through the coastal plain alluvium, a softer substrate that defies all expectation for topographic drama. Added to this ignominy is the geography of poverty within which the creek flows, a convergence that

Figure 4.9. Streambank restoration in Cobbs Creek serves as a demonstration of the ecological and recreational value of the Philadelphia region's more than 80 miles of internal waterways.

causes Cobbs Creek to lag far behind other Philadelphia waterways in terms of public appeal. The creek's ecological restoration intends to correct this imbalance. But a larger role will be played by the proposed addition of 2.7 miles of recreational trails, completing the 8.5-mile run between City Line Avenue and the John Heinz National Wildlife Refuge. To the communities of Upper Darby, Overbrook, Kingsessing, Yeadon, and Eastwick, the Cobbs Creek Greenway represents a lifeline to a natural remnant every bit as compelling as the Piedmont woods of the Wissahickon. Upon completion, thousands of residents will live within a half-mile of the Cobbs Creek Greenway, fulfilling one of GreenPlan's major objectives.

Active Mobility
In addition to the literal greening of its urban landscape, Philadelphia is also retrofitting public spaces in support of active mobility. GreenPlan Philadelphia targets 100 percent of the population to live within a half-mile of a recreational or commuting greenway. A recently finished Complete Streets manual provides a vehicle for implementation. In time it will ensure that getting to a greenway—or getting anywhere in Philadelphia, for that matter—is available to all comers whether on foot, wheelchair, bicycle, automobile, or public transit.

The nation's Complete Streets program is the result of a four-decade effort intended to legitimize cycling as a form of everyday transportation. In 2003 a national coalition was formed by advocacy groups, including the AARP, American Society of Landscape Architects, American Planning Association, American Heart Association, and Institute of Transportation Engineers, to establish policies and design guidelines in support of streets that are traffic-calmed, pedestrian-friendly, transit-oriented, and, especially, inclusive of cyclists; or, to put it in social and environmental terms: streets that promote personal fitness, that accommodate the needs of individuals with disabilities and the aged, and that, in doing so, help curb the emission of greenhouse gases and their ill effects on public health.

Philadelphia bicycle ridership grew by 151 percent in the first decade of the millennium, achieving a 2.16 percent mode-sharing (percent of commuters who bike to work as compared to other means). In recognition of the growing population of bicycle riders, the city in 2010 commissioned

Figure 4.10. Spruce and Pine streets in Center City Philadelphia lost a driving lane in favor of a buffered bycicle lane. Combined, these streets provide an active mobility connection between the Delaware and Schuylkill Rivers.

the preparation of a comprehensive Bicycle and Pedestrian Plan (BPP). Led by the Toole Design Group, the BPP proposes a tripling in commuting cycling, or 6.5 percent mode-sharing, by the year 2020. Such a target exceeds Portland's current mode-sharing percentage, the highest among American cities.

Much of this expansion is focused on the city's central area, no small task given Philadelphia's colonial pattern of narrow streets and nineteenth-century housing that places a premium on on-street parking. As a pilot project, bicycle lanes were inserted into two downtown streets, Spruce and Pine, at the expense of a vehicular lane. What were three lanes devoted to vehicular traffic—two moving lanes and one parking aisle—became one moving lane, a parking aisle, and a nine-foot buffered bicycle lane (see figure 4.10). Because of the enhanced biking space, Spruce and Pine Streets have drawn riders from other, more congested Center City streets, improving safety and increasing ridership. Motorists, too, have benefited as the single vehicular lane affords more orderly and stress-free traffic, without a measurable loss in capacity. More important, Spruce and Pine Streets now provide safe and direct access to and from the Schuylkill River Trail, Philadelphia's most prized greenway. A similar north-south couplet became operational two years later, and cycle tracks have been tested on Market Street and JFK Boulevard, two major Center City arteries. These efforts are meant to certify that losing auto-dedicated acreage in dense urban areas makes for a better and healthier environment.

Across from Philadelphia City Hall on 15th Street is the Triune, a sculpture resembling a tri-lobed Mobius strip, designed by Robert Engman and installed in 1975. Representing the seamless convergence of people, government, and industry, the work occupies an undistinguished spit of concrete well in the shadow of the massive Second Empire municipal building. Smaller in scale than the famous Claes Oldenburg Clothespin, located a block away, the Triune is a comparatively forgotten work of art. But perhaps this will change. In its green-hued topological contours, the Triune appears to embody Philadelphia's budding inside-out conversion of two centuries worth of industrial infrastructure into a greener equivalent far closer in form and spirit to William Penn's vision of a "Greene Country Towne."

Chapter 5. From Green to White:
Ecology as a Design Ethic

> *Every portion of matter can be thought of as a garden full of plants, or as a pond full of fish. But every branch of the plant, every part of the animal, and every drop of its vital fluids, is another such garden, or another such pool. Thus there is no uncultivated ground in the universe; nothing barren, nothing dead.*
> —G.W. Leibniz, *Monadology*, 1714

GreenPlan Philadelphia qualifies as an environmental initiative by any planning and design standard. Given its focus on a working nature to bring about environmental change, it is reasonable to regard it as "ecological." After all, the potential addition of trees in the thousands and the creation of miles of riparian habitat as part of creek restoration efforts would increase biological activity in Philadelphia to the benefit of people and wildlife. McHarg would certainly have subscribed to this view. But it would be wrong to regard ecology strictly in environmental terms.

Ecology is the study of relationships—of how things affect one another through interactive processes and how they transmute or evolve as a result. How people interact or view the natural world, and what changes arise from this becomes, therefore, central to any ecological discourse. This is especially so regarding design of cities, for what are cities if not an expression of a people's relationship to nature, real, imagined, or longed for? If so, we must accept that the human inclination toward abstraction, metaphor, allegory, folly, and fantasy—the work of the mind—is part and parcel of an urban ecology. These, ultimately, are the constructs by which the physical world acquires emotional and spiritual currency. If nature is demystified and its salving influence turned to the grit of cities—to the physical repository of human dreams and imagination par excellence—then ecology itself must be reframed, not merely as an approach to design but also as a foundational ethic by which to consider the future of cities.

The word "ecology" comes from the Greek *oikos* or "house." The term literally means "the study of the house," the place we live in. Most everyone involved in the planning and design of cities equates "house" with the totality of the physical world therein: all biota and associated habitat. But if we hold true that the work of the mind is part of our urban house, then we must recognize that it consists of two distinct but connected floors: one concrete, the other abstract, each existing in support of the other in a fluid exchange. (Embracing abstraction in design begs the question of art as an integral condition of urban life. Views on this question are discussed in Chapter 7.)

The notion that that which is concrete (body) and that which is abstract (mind) comprise a fluid continuum has Western roots in the philosophy of G.W. Leibniz. Aspects of his philosophy are invoked here to suggest how such a continuum could—and should—become operative in the urban realm. Given the era in which his philosophy emerged, also invoked is seventeenth century Baroque art, urban design and, architecture, and the manner in which such works embody the mind-body continuum.

It might be tempting to ignore a discussion on such an arcane matter in the context of today's planning and design. However, a modern day reading of Leibniz and the Baroque looms large as a guide toward making the artifice of cities transcendent, a prerequisite, this author would argue, towards overcoming the transcendence of "nature" as a condition of the "American Dream," (i.e., suburbia).

The foundation for this argument is derived from Gilles Deleuze's *The Fold: Leibniz and the Baroque*. Published in 1988, the work examines the Baroque through the Leibnizian lens, positing the continuum between body and mind as an eminently ecological construct (appropriately, Leibniz is regarded therein as the "father of ecology"). A discussion of both the Baroque and Leibniz is therefore germane to presenting a post-environmental, post-McHargian view of ecology by which to envision urban areas. Two key concepts are advanced: that an urban ecology should produce a fusion of the concrete and the abstract as a vehicle for green design, and that, to do so, traditional disciplinary boundaries should be blurred if not altogether erased. A trip through Leibnizian space into Roma Barroca is a fair place to start.

Body and Soul of the Baroque

For anyone with a healthy curiosity in Baroque art and architecture, the approximately two-kilometer walk from Piazza Navona to Palazzo Bar-

From Green to White: Ecology as a Design Ethic

berini through Rome's historic maze of narrow streets can be very rewarding. For someone predisposed to experience the improbable evanescence of mass under the play of geometry and light, the walk can induce sheer elation. Through a collection of paintings, sculptures, buildings and urban spaces, the walk also affords the purest possible manifestation of Leibniz' philosophy as interpreted by Gilles Deleuze, namely the indivisibility between body and mind as a fundamental condition of building the world.

The walk begins in the shadow of Sant'Agnese in Agone, the piazza's standout basilica (see figure 5.1). Designed by Francesco Borromini, the building's concave and recessed façade is flanked by tiered, seemingly twisting towers. The gap between the towers draws the eye out of the plaza toward the basilica's dome, a powerful visual release within an otherwise enclosed urban space. The eye, in essence, is forced to rove past the plaza's medieval walls to the rising figure of the dome, presaging the ultimate embrace of the human spirit by an omniscient mind. Facing the basilica at the center of the piazza is Gian Lorenzo Bernini's *Fontana dei Quattro Fiumi*, a sculpture so fluid that it hardly needs water to evoke the power of the Nile, Ganges, Danube, and Plate Rivers (see figure 5.2). Here, travertine

Figure 5.1. The dome of Sant'Agenese in Agone looms over Piazza Navona, drawing the eye outward and skyward.

has been liquefied. Four imposing human figures guard each river's point of actual watery release. Their muscular physiques, partially covered by flowing robes, yield an unrelenting plasticity. Material fluidity is a hallmark of the Baroque—it evokes, Heisenberg-like, the ever present passage from

Figure 5.2. Bernini's fountain of the four rivers exalts material fluidity, symbolizing, as seen in this image, the Ganges River and, in the sculpted figure, Asia as the continent over which it flows.

one state of being (the body) to another (the mind); that is, the passage from one floor of the ecological house to the other.

A block away to the east of Bernini's fountain is the church of Sant'Yvo della Sapienza, another work by Borromini. Through tiers of alternating concave and convex cornices, Sant'Yvo's exterior progressively diminishes as it rises to a swirling, sky-tingling top. In the building's interior, the play of curves is magnified by means of a hexagonal dome-supporting ledge that seemingly holds in judgment the tension between divine knowledge (the floor above) and the bounds of the human mind (the floor below). Exiting Sant'Yvo and past the Pantheon in the Via del Caravita is the Chiesa de Sant'Ignazio, home to one of the world's most arresting frescos. Occupying the ceiling of the central nave, the outsized tromp d'oeil depicts the work of Saint Ignatius and his exalted passage into Paradise. The painting builds upon the line work and luminosity of the nave windows, a fusion of real (body) and imagined light (spirit), causing a sense of immersion in an otherworldly landscape. But the church is perhaps more famous for its urban forecourt, a small plaza framed by buildings that inscribe in their façades a trio of elliptical arcs (see figure 5.3). The bowing façade of the central building gives the plaza spatial breadth while the smaller ellipses at either side frame the passage into it from the chaotic pattern of surrounding medieval streets. It is as if a divine chord had rung from above to drown out the din of human toil.

Beyond this matchless plaza, the walk takes us across Via del Corso up a rise to the Piazza del Quirinale to Via del Quirinale, site of two of the Baroque's most graceful architectural works: Bernini's Sant'Andrea al Quirinale and Borromini's San Carlo Alle Quatro Fontane. Sant'Andrea's façade thrusts the interior of the building onto the street by means of a semi-circular portico, the roof of which literally unfolds from the façade itself. Inside, the elliptical dome is held aloft by the ten apses that ring the chapel. Voids rather than mass appear to carry out the heavy lifting, an effect that seems to reverse the pull of gravity. Over the altar, painting, sculpture, and architecture become one, with cherubs leaping out of the canvas to materialize as sculpted figures rising into heaven through an oval oculus (see figure 5.4).

But if material disintegration is merely present in Sant'Andrea, it overflows in nearby San Carlo. The façade of the Borromini chapel is sinuous

Figure 5.3. The passage into the Piazza is defined by elipitcal cornices from flanking buldings—exalting gestures that stand in sharp contrast to the irregular pattern of streets that define much of Rome's historic (and formerly pagan) core. The piazza sets the stage for the baroque Chiesa de Sant'Ignazio.

From Green to White: Ecology as a Design Ethic

Figure 5.4. In Sant'Andrea al Quirinale, painting, sculpture and architecture become one, flowing toward the occulus in seeming opposition to gravity.

Figure 5.5. Borromini's master façade anchors a street corner. Its curving plane and virtually detached columns invite passage to the interior, augmenting the anticipation of spiritual uplift.

From Green to White: Ecology as a Design Ethic

and crenellated, a tour de force in the virtual dissolution of mass. San Carlo anchors a street corner, affording a diagonal view that magnifies the play between concave and convex surfaces and the resulting dance between shadow and light (see figure 5.5). As frozen music goes (to borrow from Goethe), San Carlo's façade packs a symphony. Inside the music continues unabated: a sinuous octagonal floor sets the stage for upward geometric clarity, first through a doubly elliptical cornice, then through an oval ring and finally through a coffered dome that pulls the eye up and outward through a brightly lit oculus. It is hard not to linger inside San Carlo and let oneself be transported by the evanescing mass into a state of seeming weightlessness. It is as if the material and spiritual qualities of lightness were one in tone and timbre.

The final stop in the walk is Palazzo Barberini, down the street from San Carlo in the Via delle Quatro Fontane. In scale and mass, the building emulates a Renaissance palace, such as Farnese, except for the wings that enclose the arrival court and the forced-perspective window reveals in the upper story of the main building. It is in the interior of the palace, however, where Baroque fluidity exerts its full force. To the right of the main entrance is a Borromini staircase leading visitors to the National Gallery of Antique Arts. It spirals upwards along an oval rail, the implied torque expressed through the twisted Doric capitals that hold the rail along its curving path (see figure 5.6). The flow of the space is dizzying, an effect that would be rather unsettling were it not for the skylight that sets the upper floor aglow, seemingly dissolving the stair into thin air.

But it is Pietro da Cortona who steals the show with the Trionfo della Divina Provvidenza, a frieze that hovers over the palace's main salon (see figure 5.7). Aside from its sheer size, luminosity, and allegorical content, the frieze embodies the indivisibility of architecture, art, and nature as an expression of humanity's earthly existence. An ornamental frame centers the composition, magnifying its upward thrust. Human figures both support and float through it—a figurative melding of the physical and psychic dimensions of human existence.

Nowhere in the world is there a remotely comparable collection of Baroque masterpieces in such close proximity to each other. More remarkable still is that all of the architectural works cited above were designed and constructed in a mere 34 years, from 1627 (Palazzo Barberini) to 1661

Figure 5.6. The Borromini stairacase in the Barberini Palace simulates a spiraling vortex, a representation of the power of the divinity to lift body and soul to the hereafter.

FROM GREEN TO WHITE: ECOLOGY AS A DESIGN ETHIC

Figure 5.7. Cortona's allegorical whirl dissolves the ceiling of the Barberini Palazzo's main hall, expressing the passage between material earth (the body) and immaterial light (the soul).

(Sant'Andrea). During the Counter-Reformation, the Catholic Church was working intently on reaffirming its dogma by means of palpable and exalted portals to the divine. To this quest, the creation of iconic works that transcended the senses through the "folding" of matter and light, solid and void, figure and story, was paramount. Formal dynamism was achieved by means of unsettled massing, geometric excess and visual illusion, employed to persuade the mind and exalt the spirit. As put by Robert Harbison in *Reflections on the Baroque*, the result is the "phantasmagoric merging of architecture, paintings and sculpture, deliberately confusing the borders between them, the goal a single overpowering effect."[1]

> If the Baroque establishes a total art or unity of the arts, it does so first of all in extension, each art tending to be prolonged into the next art, which exceeds the one before. The Baroque often confines paintings to retables, but it does so because the painting exceeds its frame and is realized in polychromatic marble sculpture; and sculpture goes beyond itself by being achieved in architecture; and in turn, architecture discovers a frame in the façade, but the frame itself becomes detached from the inside, and establishes relations with the surroundings so as to realize architecture in city planning. From one end of the chain to the other, the painter has become urban designer. We witness the prodigious development of a continuity in the arts, in breadth and extension: an interlocking of frames of which each is exceeded by a matter that moves through it.[2]

This statement, from Deleuze, offers a critical window as to the future of cities in two fundamental ways: first, through the call for a "unity of arts" by which cities should be planned and designed; and, second, through the identification of an embedded "matter that moves through it," or the substantive continuity that binds a city's story (the abstract) to its fractal array of solids and voids (the concrete). In the context of sustainable urban environments, green infrastructure must be considered the element of substantive continuity—the water, air, energy, carbon, biomass, habitat, and art that *move through* the folds of building and landscape.

From Green to White: Ecology as a Design Ethic

Requests for Proposals mandating attention to sustainable design through green infrastructure abound within professional practice. But few require the reading of the urban landscape as a state of the collective mind. Little if any value-added is accorded to the memory of a bygone place, a vision of a mythic wilderness, or the allegorical expression of human action in the landscape as a source of design. And yet every place is suffused with stories, lore, allegory, myth, and fantasy; this is a form of bedrock as real as ore ready to be mined. Mining this bedrock is an exercise in ecological design—of establishing and giving form to a place-specific mind-body continuum. It is worth pondering what city form might arise from such an exercise; that is, how abstraction can become a fold within the "matter that moves thought it." An extreme departure from Rome is necessary to posit an answer.

Fargo 365

Route 46 in North Dakota was referenced earlier as part of the national grid lore. Part of the roadway traverses prairie pothole region, a vast and unique North American landscape (see figure 5.8). It is apt, then, to turn to an urban design proposal for a downtown block in Fargo, North Dakota, to demonstrate how urban development can be rendered ecological in the Leibnizian sense. The proposal in question—for a site a mere 17 miles

Figure 5.8. In North America the prairie pothole region is the main source of sustenance for migratory waterfowl.

north of Route 46—is the winning response to a 2009 national design competition sponsored by the Kilbourne Group, a local development company intending to "facilitate a progressive urban solution through urban ideas that are viable, livable, sustainable, and beautiful." The core of downtown Fargo encompasses a mix of commercial, governmental, educational, and cultural facilities in a half-mile square adjoining the Red River, the state line with Minnesota. The tallest structure is a 20-story hotel, distinctive only for its clear rise, modest as this is, over other downtown buildings. Many downtown parcels, including much of the competition block, are used as surface parking. Given such unexceptional quality, it is noteworthy that globally-scaled ecological aspirations would be focused upon a two-acre brownfield site in the small downtown of a remote Midwestern city.

Aiming to underscore the year-round use of outdoor space in a region noted for harsh winters, the competition designers labeled their entry *Fargo 365*. (The competition entry was authored by a group of WRT designers: David Witham, Doug Meehan, Anna Ishii and Hannah Mattheus Kairys.) To meet the objective, a proposed public plaza tilts upwards from street level to a building rooftop as if reaching toward the sun, with new building massing providing protection from cold winds. In doing so, the traditional boundaries between building and landscape become diffused, as does the distinction between what is private interior space and what is public exterior space. The design coheres, Baroque-like, as a dynamic agglomeration of volumes and surfaces, solids and voids where the beginning of one transitions seamlessly into the end of another (see figures 5.9 and 5.10).

Answering the call for sustainability, the design integrates a variety of green infrastructure measures, including geothermal heat pumps, vegetated roofs and walls, photovoltaic panels, solar fabric on building facades, and vertical-axis wind turbines. But of equal if not greater importance to the project's sustainable measures is the integration of a landscape abstraction as part of the "matter that moves through it." The following is quoted from the competition entry:

> The surface of the public space recalls the iconic prairie potholes landscape of North Dakota, with the pavement opening up to hold water, plantings, and fountains in the summer, and exposing

From Green to White: Ecology as a Design Ethic

Figure 5.9. The Fargo 365 competition entry creates a "prairie pothole" at the block scale—a source of civic sustenance for Downtown Fargo.

Figure 5.10. Private and public place is conflated to underscore the overarching function of the pothole abstraction.

thermal vents heated by the geothermal system in the winter. The sloping forms create places for sun bathing, picnicking, and watching performances in the warmer months. In the winter months, those slopes become perfect for sledding, or catching a few rays of warm mid-day sun. Just as its natural counterpart serves as the center of life on the prairie, Fargo 365 envisions a prairie pothole at the center of life in downtown Fargo, a place to gather, play, watch, shop, dine and celebrate (see figure 5.11).

From the air, the real prairie potholes look like a loosely shaped assemblage of coffered wetlands. They were formed by the coarse mechanics of advancing and retreating glaciers during the last ice age. Comprising about 64 million acres mostly within the Dakotas, southern Minnesota, and northern Iowa, the pothole region is home to more than 50 percent of North American waterfowl. To this population as well as to migrating birds galore, the potholes provide an essential ecological service. Likewise, the abstracted potholes in Fargo 365 provide an ecological service, namely microclimate mitigation. But as a design feature, they are also imbued with

Figure 5.11. One "prairie pothole" is envisioned as interactive water features in summertime, and warming steam vents in winter, magnifying the site's function as a civic "watering hole" through the seasons.

From Green to White: Ecology as a Design Ethic

the distant echoes of a once abounding wilderness, a place where pioneering lives took root and flourished amid natural abundance. In essence, the competition design embraces the prairie potholes as both a poetic and utilitarian reminder of the regional ecology. The poetic dimension amply justifies the assertion in *The Fold* that "habitat includes conceptual virtue."[3]

From a design standpoint, "conceptual virtue" carries a significant burden, namely having to determine an appropriate conceptual course—having to choose as a source for design one or more among many strata of a locale's abstract bedrock.

In the context of the Fargo Urban Block Competition, for example, why would the prairie potholes be any more important to recall—to compose with—than, say, the desolate linearity of Route 46? Or the agricultural industry that economically supports the region, much of it thriving at the expense of the pothole wetlands? What makes one "possibility of composition" more fit than any other? Localism and public art form part of the answer: localism as a way to understand place-specific lore, and art as the poetic sensibility by which to choose one or more as a source of design over others. These subjects are discussed in later chapters. For now, as demonstrated by Fargo 365, it suffices to recognize the value of metaphor, allegory, and fantasy as part and parcel of a systemic approach to the design of sustainable urban environments. Key to such value is the act of composition, of injecting abstraction into the material content of a development program. A further reading of Leibniz's philosophy is useful in support of this notion.

Leibniz and Compossibility

Born in 1646, Gottfried Wilhelm Leibniz became the Baroque era's preeminent polymath. Among many accomplishments, he became famous for the development of infinitesimal calculus. Philosophically, Leibniz viewed the universe as the agglomeration of indivisible substances, or monads, each a singular reality comprising an indivisible physical and conceptual signature. Monads can be regarded as *unitary distinctions*: the physical and conceptual differences by which the world can be perceived and understood. Such distinctions are scalar, in the sense that the universe, as a monad, can be broken down in ever smaller sets, reaching well past the sphere of atomic particles. From the beginning of nothing to the summation of everything, monads represent the *ultimate continuum of infinitesimal differences* (i.e., infinity). Calculus for Leibniz was the means by which infinity gained material as well as metaphysical clarity.

Although singular in character, monads are not inactive entities; rather, they are apt to transmute into new unitary distinctions. As Ernst Cassirer explains, "the monad 'is' only insofar as it is active, and its activity consists of a continuous transition from one state to another as it produces these states out of itself in unceasing succession."[4] It is, after all, the possibility of any one unitary distinction to become something else—to *compose* itself with another distinction—that maintains the universe and keeps life itself in motion. To be compossible, therefore, every monad has to contain within itself a blueprint for the universe as a whole, past and future. Leibniz considered monads to be "pregnant" with the future and "*laden*" with the past. This is the logic by which he asserted that our world is the "*best of all possible worlds*," meaning that everything that is available (possible) for composition must already exist.

With the above in mind, ecology can be more precisely defined as the science of composition, meaning the study of unitary distinctions and the processes (i.e., relationships) by which they are constituted, sustained, or transmuted. McHarg's "layer cake" method for dissecting a landscape approaches this definition, albeit applied to a single floor of our ecological house: the body. The Layer Cake, for example, makes explicit the relationship of a Piedmont beech-maple forest to the underlayment of metamorphic schist and quartzite; or the relationship of the hydrologic cycle to the creation of sinkholes in a limestone valley.

Once the process of "composition" (i.e., nature-creation) was understood, then the impact upon it of a development program could be determined. In McHarg's pedagogy, however, aspects of the mind —abstraction—was abjured vehemently. Leibniz never entered into the design discourse, or the idea of compossibility as a way to enrich the landscape with lore, metaphor, folly, or fantasy. One is compelled to believe that to McHarg's mind the idea of "possibility" had too contingent a connotation, too much of a "what if" flavor to support his deterministic world view.

McHarg believed in the unalterable truth of cause and effect relationships, based on the physical laws of nature. Anything else he deemed to be gratuitous, arbitrary aestheticism. He taught students to understand causality as a basis for planning and design, a method aptly applicable to pristine or lightly treaded natural sites. Built-up, denser urban areas, however, comprise an altogether different landscape. Cities encompass the ever-shifting ground of culture in addition to physical factors such as geology, hydrology, soils, vegetation, and wildlife. Working in the urban landscape requires entry into the second floor of our ecological house—the biased confines of the mind as shaped by culture and community-building processes. In this realm, sifting through a universe of compossible entities before arriving at a design solution becomes mandatory. To this end, entering the design process with an open, absorbing mind becomes key; that is, a willingness to engage within any locale the widest possible field of composition, and the acceptance of informed subjectivity as a way to sift through this field and arrive at a fitting response.

An ecological ethic, then, requires an "open canvas" approach to design, one in which a site's compossible context must be discovered, examined, qualified, and in some instances tested without pre-conception for possible integration with a development program. A priori conditions, rigid canons of design, or imposed rules are in principle antithetical to such a process. To be sure, it is inevitable that some sort of apriori condition will enter into play in the planning and design of urban areas. Buildings, after all, need to answer to proven floor plate and constructability efficiencies, LEED certification criteria, established zoning, and, often times, predetermined stylistic preferences. Then again, all of these conditions exist as part of a site's universe of compossible distinctions. When a design succeeds under an "open canvas" approach, it is not because the program has met preexisting rules and regulations in a cleverer way, but because it has succeeded in measurably affecting them, changing, in effect, their apriori standing.

Because of their inherently unrestricted potentiality, design competitions are often the best way to affect established design methods and approaches. Fargo 365 stands as a unique composition, looking nothing like a rule-bound design. Like most winning competition entries, however, it is not destined to be realized exactly as envisioned. A new design team has been retained to advance the redevelopment effort, making changes to the building-landscape matrix inevitable. Still, design competitions prompt the imagination and in time cause the practice of city planning and design to shift, suggesting new models by which to construct the world. In Fargo 365 the new model points to a folding of green infrastructure and local abstraction as a way to build denser, greener, and more locally rooted urban environments—a way, in effect, to build urban America.

Injecting abstraction into the ecological discourse was not a part of McHarg's pedagogy. It is a curious omission, for the very idea of distinction as a way to understand and construct the world cannot be more of an ecological condition; that is, to understand what a place is made of and why, and to draw inspiration directly from it to compose a new possible and graspable reality.

A "Unity of the Arts"

Now to Deleuze's other key observation about the Baroque: a "unity of the arts." In seventeenth century Rome such unity was applied to the plastic arts, urban design, and architecture as a way to codify the imperative of the day, namely the reaffirmation of Catholic dogma over rising Protestantism. From the surface of the canvas to the gold-leafed finish of building interiors, the story of human redemption and passage toward everlasting life through divine guidance was made profoundly palpable. Now, if it is accepted that the creation of a healthier, more meaningful urban environment is the imperative of our time, then an equivalent unity among the city-building disciplines becomes essential. Such undertaking has vital moral overtones, for there cannot be a greater human enterprise than caring for our kind, and that of other life forms, through the world we choose to construct.

From a physical standpoint, regional planning, urban design, architecture, and landscape architecture are the principal world-building disciplines. Firms that provide services in two or more of these disciplines approach the "unity" ideal. This was the impulse that brought together urban planner David Wallace, landscape architects Ian McHarg and William Roberts, and architect Thomas Todd to establish WMRT in 1963. They believed that the task of intervening in the urban realm was too complex for any one discipline to tackle. A closed collaboration among disciplines had to be exercised at all levels of design, they believed—and indeed it was,

From Green to White: Ecology as a Design Ethic

such as in the formulation of *The Plan for the Valleys*. But multi-disciplinarity is not the same as inter-disciplinarity or even trans-disciplinarity, prefixes that more accurately convey a sense of fusion among disciplines. To be sure, clients are rarely interested in disciplinary fusion. Private or public entities invariably list explicitly the disciplines that are deemed necessary to carry out the work assignment. This practice abets the formation of teams comprising many consultants with narrowly defined areas of expertise. Lost in this approach, however, is the potential for synergistic clarity; that is, that an end result be greater than the sum of its parts.

The absence of such clarity is in part the result of professional planning or design programs that must channel students toward narrowly focused, accreditation-enabling curricula. While landscape architecture, city planning, urban design, and architecture degree programs may touch on the foundations of each of the other professions, none require students to gain competence in the gamut of factors affecting the quality of urban environments; that is: practical knowledge in building typologies, land use policy, environmental law, waste-management, vegetated roofs, urban agriculture, best management practices, urban forestry, energy conservation, material recycling, toxic remediation, community engagement, public health, and public art, among other areas of expertise. As a result, the integration of such competence as a basis for design is at best haphazard. Students with an interest in such integration take elective coursework outside their degree requirements or opt for dual degrees, say in architecture and landscape architecture. Without a comprehensive, fluid, and hybridized approach to education and practice, achieving, as mentioned above, a "substantive continuity that binds a city's story to its fractal array of solids and voids" (i.e., ecologically-based urban densification), will be destined to fail.

Within a compartmentalized disciplinary context, however, the emergence of landscape urbanism as a distinctive, hybridized approach to the planning and design of cities represents a significant breakthrough. The term came to my attention in 1998 via Carol Burns, an architect and adjunct faculty at the Harvard School of Design. We had met to discuss teaming for the master planning of the Boston Central Artery Cover (today the Rose Kennedy Greenway), which was to function as the rooftop to a one-mile depressed section of Interstate 93 in Downtown Boston. My interest in the "Big Dig" stemmed from a desire to explore a different way to think about urban space—not as a traditional green counterpoint to the built-up city but rather as a way to fuse ecology, art, urban infrastructure, and cultural heritage as a public landscape, fully integrated with new or existing development. Burns explained that such an approach sounded a lot like Landscape Urbanism, a disciplinary fusion about which a conference had been held a year earlier at the University of Illinois, Chicago. (The conference was organized by Charles Waldheim, who currently chairs the landscape program at the Harvard School of Design. Waldheim is credited with coining the term.)

Among the conference participants was James Corner, who would become chair of the Department of Landscape Architecture and Regional Planning at the University of Pennsylvania in 2000. Landscape Urbanism as a design approach to urban areas figured prominently during his tenure as chair. Corner himself is regarded as one of its early and most distinguished practitioners. Both Waldheim and Corner have written seminal articles on the subject.[5]

Our team fell short in the Central Artery master planning effort, but the idea of a holistic approach to urban design based on an infrastructural approach to the landscape persisted, prompting me to write an article on the subject. Titled simply "Landscape Urbanism," the article called for the integration of "infrastructure, building, landscape, and social functions, engaging the full gamut of spatial typologies—big and small—from gardens to rooftops to sidewalks to streets, parks, and waterways into a green meta-structure that can enhance the economic and environmental viability of urban areas."[6]

Dean Almy, director of the urban design program at the University of Texas in Austin and a leading proponent of Landscape Urbanism, regards such integration as an "alchemical relationship, one that is transformative in a tactical sense, capable of engaging the vast marginalized landscapes of the twenty-first century American City, its fringes, its underutilized interstices, its patchwork of neglected and/or denatured 'open spaces' left by development, and poorly considered law and infrastructural construction in a way that neither Landscape Architecture nor Architecture nor Urban Planning can."[7]

Almy's statement elegantly calls for a "unity of the arts," in as much as the foci of the referenced disciplines—plus others—must merge to effectively

From Green to White: Ecology as a Design Ethic

address the quality and form of cities. The use of the word "alchemy" also cleanly suggests the latent power of expansive thinking, of engaging the totality of urban resources as a mixing ground with which to effect change.

Regrettably, the notion of disciplinary unity has created more of a semantic than an applied alchemy. "Eco," "ecological," "green," "bio," and, of course, "sustainable" have since followed "landscape" as competing prefixes. "Emo" urbanism has emerged as well, with a view to suffuse the ethos of sustainability with emotional or poetic content.[8] The emergence of multiple "isms" defining similar approaches to sustainable development points to the search for a unifying planning and design approach.

"Ecological," of course, is the correct prefix; it is, in the end, the most inclusive way to conceive of the urban realm. Of interest, then, is *Ecological Urbanism*, a book edited by Mohsen Mostafavi, Dean of the Harvard School of Design. The work compiles essays from 150 contributors, among them historians, geographers, artists, scientists, lawyers, economists, engineers, and sociologists to go along with planners, urban designers, architects, and landscape architects. Issues of religion, spirituality, music, art, and literature go hand in hand with energy and water conservation, urban farming, biomimicry, and social justice. Such a mix of views and issues underscores the array of subjects, both concrete and abstract, that could (and should) be folded into the planning and design of cities. Stated in bold text is the following:

> Religion conceives of human flourishing in broad, interconnected terms that include spirit as well as the body, spaces as well as forms, and not only green but all the colors of the rainbow—a symbol of hope, expectation, aspiration and promise.[9]

Implied in this statement is the unifying inclusivity of whiteness, arguably the true hue of our ecological house. White embodies everything—the abstraction of environmentalism as much as its green practicality. White should be the metaphorical hue by which the planning and design professions build the world. But standing in the way is more than a century's worth of academic and disciplinary fortification, both in academia and the bureaucracies that support them through mandated accreditation.

The professional associations related to architecture, landscape architecture, and planning have too much vested in membership programs and licensure requirements to willingly fuse with one another. If a "unity of the arts" is to be achieved in the interest of *ecological* sustainability, it will have to be driven by an internal, profession-driven adoption of ecology as a compossible-based design ethic. Ultimately, ecology must be regarded as marrow of environmentalism, not merely its scientific lifeblood.

Chapter 6. Localism:
A Participatory Ecology

> *The intersection of nature, culture, history, and ideology form the ground on which we stand—our land, our place, the local.*
> —Lucy R. Lippard, *The Lure of the Local: Senses of Place in a Multicentered Society*, 1997

If the potholes of North Dakota are compossible entities, why not also North Dakotans themselves? But which North Dakotans? In *American Nations*, Colin Woodard catalogues the eleven "nations" of North America as derived from historical patterns of colonization and migration. Three of them come together in North Dakota, dividing the state in broad north-south bands: Yankeedom, the Midlands, and the Far West. "Yankees," claims Woodard, "have sought to build a more perfect society here on Earth through social engineering, relatively extensive citizen involvement in the political process, and the aggressive assimilation of foreigners." By contrast, the "Midlands" are Quaker in spirit, a pluralistic culture where "ethnic and ideological purity have never been a priority, government has been seen as an unwelcome intrusion, and political opinion has been moderate, even apathetic." To complete the state, Woodard qualifies the "Far West" as the "only nation where environmental factors truly trumped ethnic ones." This condition, he posits, engendered in the Far West settlers an ethos of both dependency and mistrust of a federal government (and abetting industrial barons) intent on rendering the land productive through massive infrastructure such as railroads, dams, and irrigation works.[1]

Fargo constitutes the edge of "Yankeedom" as mapped by Woodward. Seventy-five miles farther west begins the Quaker-minded Midlands, prairie pothole country. In the midst of this region is Gackle, population 350 and self-proclaimed "duck hunting capital of the world." Beyond this small but proud settlement the land transitions to Far West country, home to

91

Bismarck, the state capital. Owing to diminishing rainfall, North Dakota's Far West band is also water impoundment country. The largest of these, Lake Sakakawea, qualifies as the third largest U.S. reservoir after Lakes Mead and Powell.

Given such distinct regional identities, one would expect community voices informing potential urban development in Fargo, Gackle, and Bismarck, respectively, to be markedly different. Community specificity is a paramount aspect of an ecological ethic; it is the condition by which overarching ideas about growth and development become individuated, molded by local needs and aspirations. McHarg's pedagogy was largely in tune with this view. Before sending students into the countryside to analyze settlement patterns, he urged them to consider "why are people there, what are they doing, where are they going?" To be sure, McHarg's sought-after answers were intended to inform the "layer cake": mappable sociocultural conditions that, together with natural factors, could reveal development fitness. As questions go, however, they remain vital in understanding the ways communities affect how their world might be shaped and developed.

Cities are coralline in character: they grow by the accretion of buildings, spaces, and infrastructure, forming over time an urban crust defined by a web of communities and organizations. It is a matter of interest to review how cities coalesce as such. Seattle, for example, from a distant logging and trading post to the epicenter of the global digital revolution, or Boston—the birthplace of Yankeedom—from a puritanical settlement with a narrow religious mission, to one of the world's major centers for education, with all the cultural diversity that entails. Looking back in time, the process of city-building may appear glacial, but at the point of progress—the exercise of civic will in real time—the transformative pace is as energized as the daily flow of nutrients in a coral reef. All the more so when public works programs aim to profoundly change the urban landscape—in Boston, where burial of the Central Artery intended to reconnect the bay with the downtown via a mile-long greenway, or in Seattle, where demolition of the Alaskan Viaduct aimed to transform the waterfront into a civic and recreational destination. These are times when civic will is exercised "en masse" through public meetings, workshops, special elections, and all manner of media agitation.

Localism: A Participatory Ecology

Localism as an "Eco-Democratic" Circumstance

Public meetings associated with planning and design initiatives are attended by people with a common interest in how their communities and city should look and function. Professionals, business owners, workers, students, teachers, government officials, homemakers, and retirees are all part of the participatory class. These are the folk that packed the Washington D.C. National Building Museum in 2002 to review and discuss the future of the Anacostia River; or that thronged by the thousands in the Philadelphia Convention Center in 2009 to review and comment on the future development of the Central Delaware River. Such meetings are common occurrences in cities large and small: in Camden, New Jersey, to determine the future of the Admiral Wilson Corridor; or northward in the 178 distinct communities within New Jersey's Burlington, Mercer, Middlesex, Union, Essex, and Hudson Counties to discuss improvements to the many Main Streets that distinguish one town from another; or across the Hudson River from Hudson County in the five boroughs, 59 community districts, and 305 neighborhoods that comprise New York City to discuss, well, almost anything; and northward to Boston where countless public meetings were held for years in connection with the Central Artery, the largest urban infrastructure project in the nation's history. The Big Dig may have generated national attention for its scope, technical complexity, and extraordinary cost, but to Bostonians it was very much a local issue, one that tested the community's sense of nature, culture, history, and ideology and, by consequence, the very ground on which it stood. Development initiatives, regardless of their magnitude, funding provenance, and national significance, invariably remain the purview of affected citizenry to review, fuss over, and ultimately approve. As Lucy Lippard puts it: "the urban ego is in fact parochial."[2]

Localism is arguably the hallmark of democracy in the United States. It is the way by which communities across the land build upon what is true and unique about a place—their place. Public hearings are normally required as part of the municipal approval process for public or private development initiatives. Such hearings are likewise required on projects involving federal funding or public resources through the mandates of the

National Environmental Protection Act. But these are not the main event in the community engagement process; rather, the spotlight shines on the many meetings preceding formal hearings where people voice their values and aspirations in an effort to influence the minds of planners and designers and their enabling patron, whether they are developers, institutions, politicians, public agencies, or private funders.

The process is far from tidy. At public meetings, the airing of conflicting views and contradictory goals are a common phenomenon. As Xavier de Souza Briggs suggests in *Democracy and Problem Solving*, collective action "benefits from divergent as well as convergent thinking, from robust and flexible mechanisms for 'getting to yes,' as well as space and rules for 'having a good fight.'"[3] There is inherent unpredictability in a "good fight," a matter of concern to anyone entering the process with a specific outcome in mind.

Many planners and designers view community engagement as a burden or, worse, a sure way for good ideas to become mush. Clients seeking singular development solutions—*theirs*—can similarly abhor the process. Such dread can pit consultant-client teams against a local community, turning community engagement into a "gauntlet" to be overcome in pursuit of a good plan or design. How should planners and designers enter such a democratic process? Can good planning and design coexist with public scrutiny? How can, in essence, localism be practiced in the interest of an ecological approach to a densifying America? These were among the questions that, in the fall of 2010, permeated the consultant selection process for the planning and design of Seattle's central waterfront.

THE SEATTLE QUESTION

Few planning and design initiatives stir civic interest or generate more local polemic than city-defining public works. In Seattle it has been the long-desired removal of the Alaskan Viaduct adjoining the city's central waterfront. The viaduct was erected in the mid-1950s to ease traffic through the downtown and speed vehicular passage to points south, principally the Port of Seattle. A 2001 earthquake damaged the structure as well as portions of an adjacent seawall. Driven in part by the specter of a deadly collapse, such as occurred with the Cypress Street Viaduct in Oakland during the 1989 Loma Prieta earthquake, plans to remove the mile-long structure and repair the seawall soon followed.

As in Boston's Big Dig, a tunnel emerged as a plausible replacement for the viaduct, but at a different location rather than directly beneath it. Attached to the transportation objectives was the recovery of the central waterfront as a world-class public space knitting downtown neighborhoods with Elliott Bay—a "lush, healthy, functioning shore with dense and vibrant development along the edge of smartly planned public space, expressing the deep values of the city we love," so promoted the Central Waterfront Coalition. Seventeen civic organizations had a stake in the project, among them Allied Arts of Seattle, the Cascade Land Conservancy, Downtown Seattle Association, People for Puget Sound, People's Waterfront Coalition, Pike Place Market Foundation, the Port of Seattle, and the Seattle Parks Foundation.

Prospective design teams understood at the outset that working with these groups and the general public would be paramount. As one of four shortlisted lead design firms, WRT proposed a four-step stakeholder/public engagement process:

1. WHAT WAS: a historical review of the central waterfront intended to align the public and consultant team with a shared understanding of the site's "physical and abstract bedrock."
2. WHAT IS: a discussion of aspirations, goals, and programmatic objectives, and their validation in light of vital technical considerations, from geotechnical conditions to cost and regulatory constraints.
3. WHAT IF: the synthetical translation of what "Was and Is" into specific design proposals intended to energize, challenge, and crystallize public opinion toward a desired design direction.
4. WHAT WILL BE: the incorporation of public commentary into a comprehensive and implementable set of planning and design recommendations.

The team's approach to public involvement was introduced twice during the selection process: first through a formal presentation to a consultant selection panel, then to the public at large at Benaroya Hall, the city's principal concert venue. The latter consisted of a brief presentation followed by a question and answer period. During a cool 2010 September evening, the hall was packed with members of Seattle's design community, civic and political organizations, activists, and the general public—about 1,300 in all. On stage, backed by an oversized screen, were five representatives from WRT, including myself as team leader, and the moderator, Daniel S. Friedman, then Dean of the University of Washington's College of the Built Environment. Following the presentation, Friedman posed the evening's first question:

> How will you address or employ or otherwise acknowledge all the plans and ideas that have been generated in the past decade; how will you balance respect of that energy and history and still bring a fresh eye to the project.

The question implied that design excellence can be seriously tested—ultimately snuffed—by the public review and approval process. This was the response:

> Philosophically, as an ethic of design, [the team] is committed to discovering the energy of the place. This is strictly an ecological position, a mindset that says that you can only design things uniquely if you enter the processes at work, both natural and cultural. We are going to understand that, and it is from that source and that source alone that innovation and creativity come from—not from some other place.

It was the briefest possible way to define ecology as something more than environmentalism; to explain that a city's multifarious public voice, whether governmental or community-based, lone or group-sourced, is a part of the universe of compossibles in the Leibnizian sense; that our team subscribed to a non-apriori design approach that regards public views and conflicting aspirations as a form of localized bedrock; that it is from mining this bedrock—from listening to what people have said and have yet to say—that design ideas spring forth and mature; and, finally, that it is the

Localism: A Participatory Ecology

close relationship between the public and the designer that ultimately renders a design both unique and uniquely appropriate. In other words, public involvement is not antithetical to design excellence but the very prerequisite for it. The answer resonated with the audience; it did not, however, win the day. Following the event the selection panel reassembled to deliberate, in the end opting for a lead designer that could "bust-through" the public engagement process by force of personality and strongly held ideas. (The commission was awarded to James Corner Field Operations, a firm that had merited recent accolades for the design of the New York City High Line.) The consultant selection was accepted without argument or equivocation. It did, however, point to the need to better define urban ecology; to rationalize, in essence, the benefits of structured, open, and unbiased community engagement in the creation of urban environments.

Obtaining the benefits in question depend not on the eye but on the ear. At public meetings, each and every voice has the potential to elicit thought and reflection, to prompt further reading and research, and to leave in evidence, once the dust settles, the elemental community aspirations by which the goal of design excellence can be measured.

Take as an example the meeting that took place in Miami's Little Havana in the early 1980s to discuss the design for a new riverfront park named after Cuba's hero of independence, José Martí. The work had advanced from a winning competition entry, and now it was the community's turn—not a jury of peers—to opine on its merits. The community meeting room in Calle Ocho reverberated with chatter as the presentation went on, but midway through, the din suddenly stopped as a frail octogenarian, clad in formal military garb complete with sash, epaulets, and a slew of medals affixed to his chest, entered the room in a wheelchair. Spirited applause broke out, followed by a mob-like diatribe against Fidel Castro and the clamoring for a free Cuba.

It was a valuable interruption, however, for once the uproar subsided, attention was refocused on the design and a proposed boat embayment adjoining the Miami River, a fortuitous symbol of the exiled community's fervent yearning for a return to the motherland. That the park site had been used as the infamous "tent city" during the Mariel boatlift magnified the meaning of the embayment, turning a modest boating facility into a diaspora's embodiment of hope. From that point on, little doubt was held about the virtues of the design and its future as a locus of community identity and civic pride. Later that evening it was learned that the man in the wheelchair was the last living general of Fulgencio Batista's army. More than certifying the validity of the design, the presence of "El General" served to imbue the future park with lasting folklore, the kind that seeps into the ground as part of an abstract bedrock in ways that cannot be obtained by force of the designer's personality alone.

Another example: the occasion in Santa Monica, California, when preliminary designs for the refurbishment of historic Palisades Park and Beach Boardwalk were being considered before elected officials and the general public. Among the features of the proposed design was the restoration of Muscle Beach, loci of the gymnasts who, in the 1950s, made physical fitness a fashionable calling long before Arnold Schwarzenegger's 1969 screen debut as Hercules. The design called for new outdoor exercise equipment (rings, parallel bars, climbing ropes, pummel drums) and a lawn for the performance of individual and group stunts.

Several members of the original cast of gymnasts, then in their seventies and still in good shape, had voiced support for the proposed design. Then, a young woman walked up to the microphone. Voice quivering, she recounted how the beach had saved her life; how access to outdoor recreation and the friendship of fitness enthusiasts had rescued her from drugs, depression, and despair; how she now felt valued as a person and as a film industry stuntwoman; and how she was now devoted to saving others from a marginal life through sunshine, fresh air, and exercise. That the proposed boardwalk facing the fitness apparatus was shaped like ocean swells mattered little to her; she cared only that the city had cared enough to honor her story of redemption by affording others a similar recourse toward health and wellness. Her poignant words cemented the notion of human theater as a guiding design principle. (A more detailed description of this work is offered in the following chapter.)

A further example: the meeting at Matthews Memorial Baptist Church in Anacostia in Washington, D.C., when the community first gathered to offer their views on the revitalization of the Anacostia River, its national parklands, and adjoining development areas. Anxiety filled the small meeting room in back of the main prayer hall. No one seemed eager to listen to a group of outside experts suggest how their community could be improved. The lead consultant began its prepared remarks by reassuring those in attendance—all African Americans—that a major goal of the Anacostia Waterfront Initiative was to improve the local economy and create jobs through new development opportunities.

A few minutes into the presentation an older man stood up and proceeded to indict the team for presuming what was good for his community. With a respectful but emphatic tone he went on to describe the long-standing environmental degradation imposed by the District upon the people of Anacostia; the neglect and abandonment residents suffered in the face of polluted river waters, leaky sewers, and toxic landfills; and that it was not jobs that the community was interested in as a first priority but rather environmental justice—access to clean air and water like the folks across the river had. No applause followed. The collective silence was far more effective in conveying the community's call to "*get it right*." Here was a case where the science of ecology—habitat restoration—aligned perfectly with its abstract counterpart—environmental fairness and equality. Implicit in this alignment was the call to manifest aspirational voices that had lay silent for more than half a century—to make a dream, in effect, concrete. That the Anacostia Waterfront Initiative ultimately championed aggressive ecological restoration measures can be sourced to this initial public meeting.

But public voices do not imply strict passive listening on the part of consultants. Localism requires engagement, to enter a community as a participating voice. Briggs argues that public engagement can "change people's preferences, change the way problems are framed, bring new resources and stakes into view, and expand the menu of options under consideration."[4] The planner's or designer's role is thus catalytic: to define a community's compossible sphere and guide the debate toward solutions that advance its ever-evolving story. This approach might be called ecological placemaking.

As summarized in figure 6.1, three distinct factors influence the process of ecological placemaking: **Program**, as determined by the client's goals, community needs, and regulatory requirements; **Site**, as determined by boundary, history, environment, and cultural context; and **Ethics**, as derived from the presiding attitudes toward development. Ethics, of course, includes the planner/designer's own beliefs and predilections.

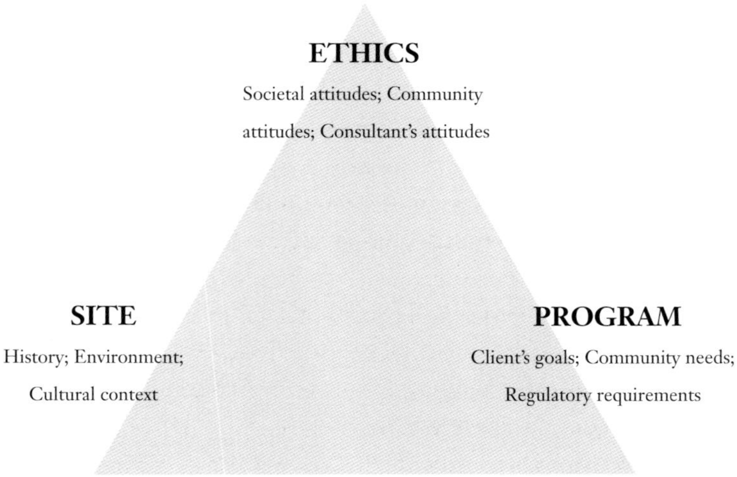

ETHICS
Societal attitudes; Community
attitudes; Consultant's attitudes

SITE
History; Environment;
Cultural context

PROGRAM
Client's goals; Community needs;
Regulatory requirements

Figure 6.1: Ecological Placemaking

The diagram in figure 6.1 serves to explain a basic tenet of localism: that a "balanced solution"—the geographic center of the triangle—does not necessarily represents the "sweet-spot" of a plan or design; that such a spot can only be derived through the process of public engagement, tilting to one apex or another depending on the best achievable convergence among program, site, and ethics. In other words, each of the three factors is a definable and interacting ecological layer suitable to composition. Unlike McHarg's "layer cake," however, the process of composition is not deterministic; it is, rather, contingent on the intersection of history, ethics, community biases, and clients' goals as points of contention within the public engagement process.

Putting this approach into practice is not a simple task. Defining for any given community a compossible sphere requires deep involvement with the local situation. As Lippard states:

> All places exist somewhere between the inside and outside views of them, the ways in which they compare to, and contrast with other places. A sense of place is a virtual immersion that depends on lived experience and a topographical intimacy that is rare today in ordinary life and in traditional educational fields. From the writ-

er's viewpoint, it demands extensive visual and historical research, a great deal of walking "in the field," contact with oral tradition and an intensive knowledge of both local multiculturalism and the broader context of multicenterdness . . .[5]

The "writer's viewpoint" is eminently transferable to that of the planner or designer: she or he must similarly understand the life of a place if meaningful participation in its ever-evolving story is to be realized. This was the operating mindset in Georgetown, Washington DC, as a consultant team prepared to interview for the design of its public waterfront which, as with Seattle's, is also backed by an elevated highway. The ecological "placemaking" diagram was shared with the consultant selection panel. The integration of art, environmentalism, and green infrastructure was offered as the ethical underpinnings of the design team. Jody Pinto, a noted artist, had joined the team in wholehearted support of this construct. It is unknown whether the diagram proved meaningful or decisive in determining the consultant selection. The question was never asked; when our team got the nod, we were simply too pleased to participate in the design of the nation's first and only national urban waterfront park.

OF WHARVES AND GARDENS

To the people of Anacostia, the "folks across the river" included the well-off residents of Georgetown, a community with a long-standing desire for a public waterfront of its own. In September 2011 the desire was fulfilled, but not before a nine-year process of public engagement and two phases of construction. Localism in this instance consisted of three interrelated spheres: the residents of Georgetown and their appointed representatives (the Georgetown Board and its waterfront committee); the District of Columbia and its various municipal departments; and the federal government and its assigned steward of public lands, the National Park Service. Each of these groups exerted influence upon the site, manifested through specific goals, regulations, and approval requirements. The district has jurisdictional control over the park's infrastructure, principally traffic, power, and drainage. As it was sited on federal land, the project also required formal review and approval from the Commission of Fine Arts (CFA), a body that is presidentially appointed to safeguard the urban quality of the nation's capital.

What Was

Georgetown was a port before it was a town. It was established in 1751 at the fall line of the Potomac River, the seam between the Piedmont and the Atlantic Coastal Plain that also marks the river's furthest navigable reach. Every major river city on the eastern seaboard, from Trenton to Richmond, was established at a similar point on their respective rivers. Slave-dependent tobacco was at first the economic driver that caused town after town to flourish at these strategic locations. Milling and tanning industries followed. In the twentieth century the Georgetown waterfront attracted a lumber yard, a cement plant, a rendering plant, and a coal-based power plant. In 1940 the Whitehurst Freeway was erected to bypass these industries en route from Arlington to the heart of the district and back, separating the town from the Potomac River. As industrial activity waned, planners were called to rethink the future of the waterfront. A comprehensive study encompassing all the lands from M Street to the Potomac River was prepared by WMRT in 1972.

Sponsored by the National Capital Planning Commission and led by David Wallace, the plan proposed the burial, Big Dig style, of the freeway to make way for townhomes along with a continuous waterfront promenade and a public plaza and marina at the foot of Wisconsin Avenue. In 1986 the land was transferred from the District of Columbia to the federal government, an act that officially designated the entirety of the site as a public waterfront. Attached to the deed was a general plan intended to guide its future development, expressly supporting the incorporation of a waterfront promenade, and a gateway plaza at the foot of Wisconsin Avenue. It would take another 16 years for Congress to designate funds for the preparation of technical design documents.

What Is

The narrow, 10-acre site extends 2,000 feet along the Potomac River between Washington Harbor and the Key Bridge. Little distinguished the place before the advent of the park, save for a segmented shoreline with a few mature trees leaning over the water and, owing to its openness, long and wide vistas downriver to the Kennedy Center for the Arts and across it to the Rosslyn, Virginia, skyline. At the outset, community residents

Localism: A Participatory Ecology

strongly supported the design of a welcoming plaza at the foot of Wisconsin Avenue; they also voiced a strong preference for a passive rather than an active recreation program. The park, in effect, had to feel and function as a water-oriented pleasure ground. Many community meetings were held before settling on the detail of the recreation program. A labyrinth soon emerged as a favored feature, a matter of some concern to National Park Service officials who feared its potential use as a locus for obscure and unwelcoming rituals. But the labyrinth stuck, as did a fountain at the main entry plaza, steps leading down to the water to view regattas, a waterfront promenade, and bikeway completing the "missing link" of a 220-mile regional trail stretching from the terminus of the C&O Canal in Cumberland, Maryland, to Mount Vernon, Virginia (see figure 6.2).

What If

In the spirit of "why are people there, what are they doing, where are they going," Jody Pinto and the design team agreed early on that the site's historic contribution to the nation's shipping and seafaring lore was an apt

Figure 6.2. The eight-acre Georgetown Waterfront Park is wedged between the Potomac River and the Whitehurst Freeway. Its dedication in Septmember 2011 culminated a long-standing desire to transform the former industrial waterfront into a national and community recreational asset.

place to start the public dialogue. The community, however, also manifested interest in the town's historic garden tradition. Dumbarton Oaks is among the famous period gardens in the United States. Located at the top of the hill not far from the waterfront, the garden contains colorful planting beds, courtyards, vine-laden pergolas, and a wilderness area. To be sure, Dumbarton Oaks is not a community resource; it is run by Harvard University as an enclosed and exclusive research facility. Still, its ornamental quality resonates with the town's well-tended yards, gardens, and pocket parks. The design challenge was thus set: to "compose" the grit of a working wharf with the bountiful greenery of a garden. The garden ideal was addressed by sectioning the park into discrete spaces, each framed by tree-lined entry paths encasing diverse groundcover plantings and terminating in overlooks from which to gather the river views. The focus on structured greenery received predictable support. But the community was divided over the expression and cost of the "wharf" components.

Jody Pinto had envisioned a series of shade structures at the waterfront overlooks. They consisted of 70-foot tilted fiberglass poles with adjoining billowy stainless steel mesh canopies, a direct allusion to the schooners that two centuries earlier had docked at the site, masts piercing the Georgetown skyline (see figure 6.3). Pinto had also envisioned a 150-foot pergola facing the river steps as a sinuous, fiberglass-topped structure evoking the ephemeral qualities of ocean waves and clouds. Overall, the design of the waterfront eschewed the town's traditional wrought iron detailing and brick paving in favor of stainless steel, cobblestones, and asphalt block pavers, materials that better approached the rugged simplicity of a wharf. Some community members resisted the austere materials palette while others supported more current design trends, especially the integration of public art such as Jody proposed for the overlooks and pergola. Georgetown, in effect, is a community with parallel narratives: one conservative, the other progressive. The National Park Service remained neutral, preferring to lay the story of the place at the doorstep of the CFA for review and consideration.

What Was (and Became)
The evocation of the site's history as a working port was argued forcefully as a foundational design concept. To press the point, a large model of the

Localism: A Participatory Ecology

Figure 6.3. During the schematic design phase, artist Jody Pinto aimed to evoke, at a proposed overlook, the town's 19th century seafaring industry through tilted fiberglass masts and billowy stainless steel mesh canopies. The proposal was rejected by the Commission of Fine Arts, citing concerns over their size and their impact upon river views (the Key Bridge is seen in the background).

park was built as well as sophisticated 3-D simulations of the overlooks. During the initial CFA review meeting, the Commission members applauded the integration of art with the park design. A joint tour of the site was conducted—a "walk in the field"—to better assess the implications of the proposed design.

The "traditionalists," however, mounted an effective campaign against the public art elements, the overlooks in particular, causing the CFA to reverse course and disapprove the overlooks and pergola. But the CFA supported every other feature of the park, including the use of stainless steel, cobblestones, and gray asphalt blocks along the promenade. In so deciding, the CFA struck a measured progressive tone: the "wharves" argument was valid, it ruled, but not enough to append to it the high drama of art. Without public art in the mix, the overlooks and pergola were redesigned. The former became tilted slabs of granite etched with historic imagery, features inspired by the metamorphic intrusions that define the Piedmont geology over which the town lies (see figure 6.4). The latter retained a sinuous form but functions strictly as a steel-framed, shade-giving vine structure (see figure 6.5).

Elsewhere the park exhibits due infrastructural greenery, namely rain gardens, hundreds of feet of bioengineered revetments, many new trees, and acres of pervious groundcover in place of hardened asphalt. At the labyrinth, a special bench built out of wood from recycled pickle casks offers a perch from which to view the waterfront. The bench was part of a gift, from a private foundation, to build the labyrinth. Below the seat is a wooden pocket containing a book and pen for people to record their thoughts and feelings. One entry reads: "Beautiful Fall Day. A formation of geese. The magical sunset. I never noticed the river flows that direction." No obscure rituals have to date taken place. (The labyrinth was funded by the TKF Foundation, of Annapolis, Maryland, in support of its mission to build "sacred spaces" in urban areas (i.e., refuges for meditation and introspection). The foundation has helped fund gardens and other labyrinths on hospital grounds, at penitentiaries, and in low-income communities.)

Figure 6.4. In lieu of Pinto's structures, the overlooks were ultimately fitted with sculpted slabs of granite etched with historic imagery of the Georgetown waterfront.

LOCALISM: A PARTICIPATORY ECOLOGY

Figure 6.5. A pergola and river steps mark the bend of the Potomac River, affording views of the Kennedy Center for the Arts and Roosevelet Island. In time, flowering vines will reach the pergola's mesh roof, melding the "garden" and "wharf" design concept.

It would be easy to qualify the design of Georgetown as a compromise. Pinto's artwork, after all, failed to be accepted. Yet it is clear beyond doubt that the act of civic engagement delivered a design that superseded people's expectations. In promoting the fusion of art, environment, and infrastructure, the team succeeded in *framing the problem in new ways*. In securing the support of a private foundation to fund the labyrinth, new resources and stakes were brought into view. And through the long and intense three-sphere community engagement process, *preferences were changed and a spirited dialogue about the town's identity was established*. A new chapter in the story of Georgetown was thus written.

The case of the Georgetown Waterfront Park unarguably supports the notion that there cannot be compromise in the advancement of localism; rather, it enables the production of the story of community as a fundamental democratic act, made real and palpable as public space. Advancing urban densification requires such a condition: people, ultimately, must inhabit places through which they can enact their dreams and aspirations, as members of a collective narrative.

Chapter 7. On Public Art

> *Whether Mr. Mutt with his own hands made this fountain or not has no importance . . . He took an ordinary article of life, placed it so that's its useful significance disappeared under the new title and point of view – created a new thought for that object. As for plumbing, that is absurd.*
> —Marcel Duchamp, from anonymous editorial commenting on a photograph of a urinal, 1917

Much like a walk in the woods can be remembered by distinctive encounters—with a fallen tree, a jutting rock, a wet patch, the fleeting sighting of wildlife—art can engender similar recall in the urban environment through iconic works that in concept and craft elicit wonderment, puzzlement or joy. Such a vital function is achieved by embracing metaphor, allegory, folly, and fantasy as site-specific conditions—*the abstract floor of the ecological house*. The idea of place, and the emotional imprint it engenders, is as much part of a locale as the rock strata beneath the surface. The capacity to imbue concreteness with the power of the abstract, and vice versa, as an art form, must unarguably be a core concern to the design of the urban environment. And who is better equipped to act upon this concern than an artist? Who better than the artist can sift through a universe of compossibles and posit choices that can exact out of a community's evolving narrative a measure of poetry? Art cannot be disassociated from habitat—and in some form or other it never has, from the millennial paintings of Altamira to Claes Oldenburg's *Paint Torch*, installed in 2011 outside the Philadelphia Academy of Fine Arts.

Since the establishment of Philadelphia's percent-for-art program in 1959—the nation's first—cities and towns across the land have endeavored to enrich the public domain through similar programs. The works vary greatly: from iconic objects such as the aforementioned *Paint Torch*, a two-story upright paint brush complete with a bench-sized blue dollop at its base; to grand civic places such as Isamu Noguchi's Bayfront Park in Miami, to entire building facades such as Ned Kahn's *Wind Veil*, a six-story, block-long sheet of fluttering panels covering a Bank of America parking garage in Charlotte, North Carolina. These and thousands of other works help make urban places distinctive and uniquely local. As stated by Americans for the Arts, "absent public art, we would be absent our human identities."[1]

Given the value of art in the creation of an "urban nature," it is surprising that so few of the major American universities offering master's programs in planning, urban design, architecture, and landscape architecture make a reference to art as part of their missions.[2] In stark contrast to the Baroque enterprise, art does not presently figure as a core concern among the city-building professions. It is uncommon for architects and landscape architects to collaborate with artists on urban projects as equal partners. Many of them regard the integration of public art a demand every bit as burdensome as the process of public engagement. Whether a sculpture, mural, light effect, digital work, interactive display, or performance space, making room for such works in buildings and landscapes is often viewed, at best, as additive and, at worst, as extraneous features that detract from or compromise the clarity of the design intent. Many public art programs abet such an antipodal relationship by imposing the integration of art on buildings and landscapes well past the initial project programming and schematic design phases. In other cases, art programs dispense with the collaboration altogether, engaging artists to create and install works after a project has been built and the design team dismissed. Art critic Jeff Kelley summarizes the frequent discord between artists and designers:

> Perhaps the most typical misunderstanding architects have about artists is that they want to build 'art' into their projects, or that they want to make the architecture itself; that is, that artists want to play at being architects. Perhaps the most typical misunderstanding artists have about architects is that they also play at being artists, mistaking formal elegance for meaning, historical reference for tradition, and an abstract equation of form-to-function for use.[3]

McHarg's regard for artists advanced well past "misunderstanding"; it approached the sphere of contempt. He believed that art had "occult and esoteric pretensions and an intrinsic obscurantism." He downplayed art as a distinct field, granting lawyers, biologists, and economists a creative potentiality equal to that of painters, poets, and playwrights.[4] To be sure, his antipathy related to art applied to the making of landscapes, specifically the derivation of arbitrary designs uninformed by ecological process that conferred upon the natural world a kind of un-fit gratuity.

McHarg's negative views on art are understandable. Land art had emerged in the mid to late 1960s, and little of its scope was ecological by any scientific standard. One of the movement's first major works was Alan Sonfist's "Time Landscape." Developed in 1965 and installed 13 years later, the work sought to recreate a native forest's successional stages, from meadow to woodland, in about a quarter-acre plot at the corner of West Broadway and Houston Street in New York City. Attempting to "freeze" stages of natural succession at a bustling New York City intersection would have been, to McHarg, absurd. Clearly, a pre-colonial ecology cannot be established within a fenced plot, zoo-like, embattled by urban noise, dust, exhaust emissions, and tossed refuse. Similarly, nothing about Walter De Maria's 1977 *Lighting Field*, a work consisting of 400 stainless steel poles spread over a two-third-square-mile grid in the New Mexico desert, constituted a benefit to the local flora and fauna.

It is curious, in retrospect, for McHarg to have had such antipathy towards land art. Much of it, after all, is concerned with the unique confluence of nature and culture as an environmental descriptor—a position that parallels an ecological understanding of the world. The work of Christo and Jeanne-Claude, for example, is hardly transferable to locations other than where they are sited. Whether saffron curtains pacing walkways in Central Park (*The Gates*) or flamingo-pink skirts encircling spoil islands in Miami's Biscayne Bay (*Surrounded Islands*), it is the specific quality of each site that gives each work visceral power. Too, works of land art have been noted for extensive public debate and involvement, in addition to extensive environmental review. In *Surrounded Islands*, for example, environmentalists protested against the project and were mollified only after extensive studies proved that the work would have little impact on the islands' population of pelicans. The process of debate and approval was widely covered by the local press, rendering the work as vivid a public city-building as that of the Seattle central waterfront, and yielding as much lore.

Today it is virtually impossible to forego public art as part of any urban planning and design discussion. There are too many precedents that prove its worth, too many artists working in the public realm, and enough municipal officials keen on facilitating its integration with the public realm. If we accept that an urban ecology engages abstraction as a condition of localism, then it follows that the sensitivities of the artist—whether painter, sculptor, performer, musician, or poet—can be instrumental in taking the measure of and expressing a locale's unique place in time. And yet public art—*art as a sensibility*—remains largely unrecognized as an essential force in the city-building enterprise.

The weary dance between artists and designers begs the question: how can the unchained sensibilities of the one meld with the function-driven tendencies of the other? Clearly, a convergence in method and approach is needed. Cultural critic Jeff Kelley offers an answer. He maintains that "the best public artists want to be partners in a creative process. It is out of that process that their work emerges and has meaning. In this sense, an artist's 'work' may be more like a verb than a noun, visible not only in space but over time." The concept of works of public art emerging "over time" spells localism: that is, the engagement of community through multiple encounters, hearing the voices at work and, through an open mind, allowing them to influence one's thinking. An open mind also must be kept between artist and designer. Kelly goes on to say that "collaboration is a process of mutual transformation in which the collaborators, and thus their common work, are in some ways changed."[5] In other words, it is not merely the meeting between a site and the artist or designer that can elicit a "local" design response, but also the meeting of the sensibilities and tendencies that cohere upon it. In essence, an urban ecology admits, between participants, the co-evolution of thought.

Site Specificity

A central tenet of public art is site specificity; that is, the notion, in support of localism, that each work answers to the unique characteristics of a place, both concrete and abstract. Sharing this tenet is obligatory as a point of departure for artist-designer collaborations. Absent an open mind with which to engage the local, the melding of artistic and design sensibilities becomes a hit-or-miss proposition. A discussion of site-specificity is presented ahead in an effort to inform the point. More important, it is offered as a way to contextualized "eco" site-specificity as the end-game to the fusion of art, localism, and green infrastructure.

Robert Irwin is among the early proponents of site-specificity as a condition of art, especially as related to the urban environment. He characterizes his approach to public art as "site conditioned/determined," where the "sculptural response draws all of its cues (reason for being) from its surroundings." As a working method he proposes an intimate "hand-on reading of the site ... sitting, watching, and walking through the site, the surrounding areas (where you will enter and exit), the city at large, or the countryside." His words echo Lucy Lippard's call to experience a place intimately to gain a true measure of its genius. Irwin goes on to affirm, as a working principle, that for any art to be conditional, it "can only occur in response to a set of specifics ... possess no transcendent criteria (truth). . . and have no grounds for predetermining (preplanning) its actions."[6] Site-specificity, therefore, mandates complete receptivity to the local as a source for invention.

Among Irwin's noted works is *Two Running Violet V Forms* (see figure 7.1). It is one of 18 site-specific works from prominent environmental artists that dot the campus of the University of California at San Diego as part of the institution's Stuart Collection. Installed in 1983, *V Forms* consists of two acutely angled panels of blue-violet plastic-coated fencing set amid a grove of eucalyptus trees. The fine mesh slices through the grid of trees 15 feet above the ground, coming into view without warning, as if a panel of La Jolla coastal sky—all too often clear and blue violet—had been cut and pasted in the middle of the grove to magnify the phenomenal quality of the landscape. Irwin refers to such play between art and land as a "confrontation . . . a rare occasion with the unsuspecting potential to 'see again'"[7]; that is, as a way for art to cause a new perception and draw out of the commonplace new possibilities for meaning. Today, the eucalyptus grove, symbol of the university's perch atop the Pacific coastal ridge, cannot be imagined without "V Form" in its midst.

Two years after Irwin's work another significant work was added to the Stuart Collection: Bruce Nauman's *Vices and Virtues*. The piece wraps the top of the university's structural systems laboratory, a six-story single vol-

Figure 7.1. Irwin's site-specific V fences *are set in a eucalyptis grove, a remnant of a former plantation that today helps define UCSD's identity as an urban campus.*

ume building dedicated to the analysis of building support components. Large multi-colored neon letters intermittently flash pairs of vices and virtues—Sloth/Charity, Anger/Fortitude, Avarice/Justice, and so on—until all 14 classic vices and virtues course through the building façade like a moving frieze. The choice of venue speaks of the glaring limitations of science to affect human nature, especially that which is used to construct the world. Through the fusion of art and architecture, matter and spirit become conflated as a condition of our existence.

SITE SPECIFIC UTILITY

It is the inherent ability to affirm or challenge our views of the world, to make us think and feel what we make of it, that distinguishes public art from the craft of design—all the more so when the world we know is undergoing rapid and widespread climate change. It is extraordinary, then, when the locus of public art is a place of refuge, a place that soothes the mind with an immersion into a seemingly pristine, less convulsed past.

In Philadelphia such a place is the Wissahickon Valley Park, a six-mile steeply wooded remnant of the Appalachian Oak Forest located in the northwest corner of the city. Hardly a glimpse of the city skyline is visible from within. Sculpted over millennia by the creek bearing the park's name, the place contains dells, escarpments, rock outcrops, sandy beaches, rills, eddies, deep waters, and stone laced rapids, all part of a meandering gorge. It contains, as well, a wide social trail, miles of rugged excursion paths, historic ruins, and a museum.

And then there is *Fingerspan*, a 60-foot meshed passage over a ravine, built out of self-rusting steel and shaped like an index finger (see figure 7.2). You come upon it along a foot path with little warning, appearing as an improbable figure at the top of a stepped rise flanked by crags of schist. The structure was airlifted into place in 1984 amid much fanfare, a major milestone in the Association for Public Art's long and storied history of commissioning art works for the benefit of the public. (Established in 1872, the Association for Public Art, formerly the Fairmount Park Art Association, is the nation's first private organization to seek the integration of art in the public realm. It differs from the Philadelphia's percent-for-art program in that the latter is a public entity, supporting the integration of

Figure 7.2. Artist Jody Pinto's Fingerspan *bridge spans a ravine in Wissahickon Valley Park. Hikers with dogs invariably pick them up to ease their fear of the the grated metal flooring. Such a prompt underscores the work's re-dignifying power between people and all living things.*

art through the redevelopment of city-controlled lands.) *Fingerspan* was designed by Jody Pinto in collaboration with engineer Sam Harris, and it triggers much thought: Is the fingertip pointing towards the ground, or merely resting on it? Why an index finger, or a finger at all? Is it a reference to that other famously extended finger gracing the ceiling of the Sistine Chapel?

As an object in a natural landscape (insofar as the park is a remnant natural forest), *Fingerspan* qualifies as land art. Inspired by its surroundings and shaped specifically in response to the local topography, the work further qualifies as site-specific art. And yet it is also utilitarian. *Fingerspan* is a foot-bridge used daily by excursion walkers and mountain bikers. Much of Pinto's sculptural work supports normal human activity—walking, sitting, gathering, viewing, being. It is driven by formal considerations to be sure, but also by deeply held concerns about suffering and healing of the human body and, by extension, of the landscape. Highly sensitive to women's issues—sexual abuse especially—Pinto, in life, mixes activism against

On Public Art

aggression and injustice with the female capacity for empathy and curative restoration. To her, a degraded landscape is a body in pain, an abused and degraded body that demands action towards physical repair and restored dignity.

Fingerspan both repairs and dignifies the Wissahickon Valley Park; it removes the erosive action of feet clambering up and down steep terrain while yielding a suspended, screened view of the landscape that magnifies its quality as a sanctuary. In doing so, art is thrust upon the quotidian as a life-affirming link between the functional and the sublime. The work was not inspired by Michelangelo's masterpiece. Pinto's intention was to place people inside the body, inside a part that is essential for building the world—the hand, the finger—and to suggest through its transparency the tenuous nature of the body but also the miracle, under such tenuousness, for the finger, the hand, the body, to make things and to provide relief to the pained landscape. This is the stuff of drama, the fusion of body and mind as transcendent experience. Or, at the very least, a very cool place to walk through and stop, for a moment, to take stock of the world around us.

Santa Monica BIG

Pinto also applied a healing ethos to the Santa Monica coastline. The work focused on the two distinct sections of the coast at either side of the Santa Monica Pier, a historic amusement park and one time terminus of famed US Route 66. To the north of the pier is Palisades Park, a Victorian-era pleasure ground at the edge of the Pacific bluffs; to the south, a beachside promenade—South Beach—backed by mid-rise housing, hotels, and a health spa. In their glory days these landscapes provided, respectively, a perch for quiet strolling and picnicking and an arena for games and frolic under the beating sun. Decades of neglect had left both places degraded. At Palisades Park, trees, trails, and fencing had been erased by the receding bluffs, lawn areas had become patchy, and restroom buildings dark and dingy. At South Beach, paving was cracked, play areas were old and uninviting, and only a few worn exercise bars remained of the formerly vibrant Muscle Beach.

Aiming to rehabilitate the place, city officials were intent on creating a seamless integration between art and design, from initial programming through to project execution. Following an unusual artist/designer selection process, in 1995 the city retained Pinto, jointly with WRT, to redesign the space. (Santa Monica's Office of Cultural Affairs instituted a two-step selection process. First, over the course of a day at the Civic Auditorium, seven short-listed artists and design firms presented their qualifications to each other in the hope of exposing potential affinities. Second, paired artists and design teams returned a month later to explain before a selection panel how the team intended to collaborate. While Pinto and I had not previously met, each of us had a reference of the other, and soon after the first event we agreed to collaborate.)

Owing to the diversity of the oceanfront and the inevitable phased implementation of any planned improvements, the project was officially labeled the "Beach Improvement Group Project" or simply BIG. Many public meetings were held to elicit ideas and concerns regarding the reconstitution of the park and boardwalk's as a coastal wonder. Some of them involved long days at the park eliciting and receiving comments from all comers. Others took place at city hall, such as the occasion referred to in the previous chapter. Ever present were the "guardians of the beach," members of the original cast of gym-

Figure 7.3. View of South Beach. What used to be a straight path between buildings and sand became a wavy promenade with walls of varying heights, a refurbished Muscle Beach, Chess Park, and a shore-themed playground.

nasts, wrestlers, and weightlifters who, from the late 1930s through the 1950s, turned the sandscape into a theater for balancing stunts and acrobatics. One of the gymnasts, Russ Saunders, had become a prized Hollywood stuntman, appearing in *Singing in the Rain* among other major motion pictures. Russ and partner Paula Unger eagerly recounted their feats and accomplishments to the design team. In youth Russ did not cut a Mr. America figure. His physique was lissome but with perfect muscle tone, a quality that earned him the nod from Salvador Dali to pose, head bowed and strapped for hours to a tilted plank, while he painted the Christ of Saint John of the Cross. It is perhaps all too appropriate that the Christ in Dali's masterpiece appears suspended over a sandy shore.

To Pinto and the design team, such personal histories were a local source of alchemy: for silica to transmute into sand, sand into beach, beach into a cultural terrain. As a "circumstance," the BIG project demanded an immersion into the ebbs and flows of daily life that turn the arc of history into a recurring human and natural spectacle. In their voices lay the historical record; in their physiques the dignified passage of time; and in their collective spirit the power to inspire designer and artist to recapture the theater of muscle and light.

At South Beach, "theater" became manifest through wave-like seating walls that edge the promenade, itself a sinuous feature recalling the wash of ocean surf. Of varying heights and widths, the walls help control the flow of wind-borne sand onto the paving while creating a stage for people to sit, stand, lean, perform, and play in multiple ways. Embracing their site-specific utility, Pinto turned the walls into works of sedimentation and erosion, a metaphor of nature's eternal performance upon the physical world. Swaths of exposed aggregate, along with embedded flotsam and jetsam, were exposed along the walls through power-washing soon after the concrete was set. In their texture the walls recall the conglomerate geology of the coastline, suggesting the process of constitution and reconstitution by which all matter settles and, temporarily—*fleetingly*—provides humanity an existential foothold (see figure 7.3).

At Palisades Park, wind and rain had sculpted the bluffs into an irregular edge, erasing sections of a few feet to over 50 feet of the conglomerate rock, along with palm trees and sections of fencing (see figure 7.4).

Figure 7.4. "Stressed" concrete seat walls mitigate sand drift to the promenade at Palisades Park. Here, the walls edge a rest facility, the roof of which was designed in collaboration with artist Jody Pinto to evoke the playful quality of beach balls.

But far greater theatre was produced by the California Incline. The historic 1930 roadway exposes a quarter-mile section of the bluffs as a sheer cut, leaving a jutting perch as it turns towards Ocean Avenue following its ascent from the Pacific Coast Highway. To the design team, this was the place where people could experience the geologic void and see the Pacific expanse from a different vantage point. A viewing platform was proposed, tilted upwards to engender a rising sensation. Pinto shaped the platform in the form of a trapezoid, the far end wider than the entry point to progressively expand the field of vision. She also clad the platform in wood and edged it with a thin rail to evoke the visual extension obtained from the deck of a ship. A 30-foot internally lit and tilted mast completes the nautical allusion. It is a perch that, without apology to the vehicles that rush by, wheels churning and headlights

On Public Art

glaring at dusk, stops the clock at the edge of the continent and invites the senses to take in the immensity of the sea and sky (see figure 7.5).

The project's success as a designer/artist collaboration became clear during the team's final presentation before the Santa Monica Art Commission. The design proposals were approved unanimously, but not before one puzzled member of the commission asked: "Where's the art?" The expectation of art as a distinguishing characteristic of a place had been superseded by the creation of the place itself, a realized "unity of the arts."

Site Specific *Ecological* Utility

The proliferation of public art programs and public artists has engendered many conceptual strands, each bearing differently upon issues of utility,

Figure 7.5. At the Palisades, a beacon rising out of a projecting deck marks the passage into Santa Monica from the California Incline.

community identity, environment, and social justice among other concerns. Site-determined, site-oriented, site-referenced, and site-conscious are among the variants, all falling under the New Genre public art label. Judith Baca's *The Great Wall of Los Angeles*, Suzanne Lacy's *Full Circle*, Mierle Laderman Ukeles' *Touch Sanitation*, and Rick Lowe's *Project Row Houses* are emblematic examples of the genre. These are works focused on community issues as the medium for art, whether it be the mindless generation of waste in New York City and the life of the sanitation workers that deal with it in relative obscurity (Ukeles), or the revitalization of a rundown neighborhood in Houston through a home renovation, art-based program (Lowe). In these and other New Genre works, the site itself as the locus for art is largely dispensed with, functioning more as a conceptual vessel.

New Genre also encompasses works that shed light on environmental issues, both as critique of social behavior (e.g., consumerism, resource depletion) and as a way to effect positive change. Environmental Art, Eco-Art or Ecoventions are terms associated with such a focus.[8]

Owing to the all-encompassing purview of ecology as the study of "our house," Eco-Art suffices as a label. Noted practitioners include Betsy Damon, Mel Chin, Buster Simpson, Lorna Jordan, Jackie Brookner, Patricia Johanson, and Jack Mackie. Their mission is clear: to highlight living systems, and promote environmental health and community well-being. By necessity, Eco-Art engages federal, state, and municipal agencies charged with monitoring and regulating environmental quality. Engaged, too are the planning agencies and community support groups with interests in the effects of environmental amelioration. Several works are worth noting in support of green infrastructure and localism as components of viable urban environments.

Mel Chin's "Revival Field"

A seminal work in the rubric of Eco-Art is conceptual artist Mel Chin's *Revival Field*, a 60-foot square test plot in a superfund site near downtown St. Paul, Minnesota. Working with Rufus Chaney of the U.S. Department of Agriculture, Chin's intent was to raise awareness of humanity's noxious waste practices by training the crosshairs of art upon a toxic landfill—while simultaneously testing natural soil remediation methods. This may be the first instance where a "working nature" was the substance of public art (ironically, for safety reasons, the site could not be publicly visited). Installed in 1990, the fenced plot was planted with sweet corn (*Zea mays*), bladder campion (*Silene cucabalis*), and other hyper-accumulator species capable of extracting heavy metals from the soil. Through such plantings Chin aimed to test the potential for phytoremediation in a virtually dead— and deadly—patch of the city. Pressing his case for funding with the National Endowment for the Arts, Chin referred to the work as "*sculpting away social problems.*"[9] More specifically, he viewed the work as

> ...a sculpture involving the reduction process, a traditional method used to carve wood or stone. Here the material being approached is unseen and the tools will be biochemistry and agriculture. The work in its most complete incarnation (after the fences are removed and the toxin-laden weeds harvested) will offer minimal visual and formal effects. For a time, an intended invisible aesthetic will exist that can be measured scientifically by the quality of a revitalized earth. Eventually that aesthetic will be revealed in the return of growth to the soil.[10]

There is hardly a better example than *Revival Field* for calling attention to the issue of environmental degradation, consumption, and waste. McHarg surely would have applauded this kind of art, one with an "invisible aesthetic," backed by the rigor of plant ecology. He may have demurred on the temporary fencing, however, a seemingly unnecessary addition. The entirety of the landfill, after all, was off-limits to the public. And yet it is the fencing and not the plants that makes the work transcendent. It consists of an enclosing outer square enclosing an inner circle divided by paths in the form of a cross, a clear allusion to the archetypal partition of the paradise garden. It is this deeply ingrained vision of blissful comfort juxtaposed with a superfund site that strikes the poignant chord. For how can we ignore the self-created toxic substrate over which our urban garden lies?

Buster Simpson's *Host Analog* and Others
About the time the sweet corn was taking root in St. Paul, artist Buster Simpson was assembling *Host Analog* in Portland, Oregon. Seattle-based

Simpson is chiefly preoccupied with the nature of civic space and the role of living systems in it. *Host Analog* consists of a fallen and bucked Douglas fir recovered from within Portland's watershed and displayed as the host for ecological succession in the forecourt to Portland's Convention Center. An overhead irrigation system provides the requisite moisture for new shoots of fir, ferns, and other forest plants to grow on or astride the sections of the original tree, all contained within a curbed planter.

But why limit Eco-Art to a particular site? Why not whole sections of a city? In 1998 Simpson was retained by the Seattle Arts Commission to prepare an art master plan for the Seattle Public Utilities. Called *Poetic Utility*, the plan guides the development of infrastructure throughout the city. It focuses on drainage, waste water, solid waste, and recovery, and air and water pollution as the media for art, spanning the regional, neighborhood, and home scales of the city.[11]

The plan calls for artists to be engaged in the design of habitat for the Chinook Salmon at one scale and individual building cisterns on the other. Simpson implemented the latter through *Beckoning Cistern*, a work that captures roof stormwater from a downtown loft building and directs it via inventive scuppers and storage tanks to street-side rain gardens (see figure 7.6). In 2011 he went on to design "Bio Boulevard and Water Molecule," placed on the median of the entryway to Seattle's new Brightwater water treatment plant and featuring an outsized pipe that distributes gray water for wetland creation. The pipe is tilted, suggesting as a first impression that the water is flowing uphill—that the struggle to render the environment clean and habitable is indeed as difficult as it is urgent, especially as related to water, a primordial substance.

Lorna Jordan's *Theater of Regeneration*

But why settle for the city scale? The idea that public art can capture the scale of an urban region is grandly envisioned by Lorna Jordan's *Theater of Regeneration*, known officially as the Broward County (Florida) Environmental Art Master Plan. The project was the result of a 2000 bond referendum that allocated $400 million to the preservation of county open land and park improvements. It establishes "an Art Network of systems, sites, and events that connect people to the environment through an experience of art. The

ON PUBLIC ART

Figure 7.6. In Beckoning Cistern, *artist Buster Simpson celebrates the use of cisterns and rain gardens as green infrastructure.*

Network traces a remarkable diversity of journeys that offers encounters with the developed and the wild; the past and the present; and the imaginary and the real . . . a system of loops and lenses revealing the ecological, social, cultural, natural, phenomenological, geographical context and historical layers of Broward County within South Florida"[12] (see figure 7.7).

Anyone familiar with the flatness and dull urban chroma of South Florida will admire Jordan's effort to engage Broward County's landscape and put it front and center as a civic trove. The plan, in effect, creates a new urban geography based on "layers" of the county's biocultural body (topography, hydrology, climate, geology, vegetation, people); "loops" of circulation linking particular points of interest (parks, nature preserves, waterways, civic destinations); and "lenses" that bring sharper focus to the cultural life of the county through historic interpretation, artwork, displays, and performances at key sites. Jordan envisioned such a focus as "art-as-place through ecological infrastructure."[13]

The allusion to ecological "layers" cannot be more closely aligned with the basis of McHarg's method for planning and design. Yet Jordan deepens the meaning of the term by associating with it the idea of "journey," by engaging the landscape not just as a way to understand it, but to recover a lost experience. In this way it is not the return of a wandering son that may become prodigal, but the return of a forgotten landscape. *Theater of Regeneration* fuses the concrete (science) with abstraction (memory). In doing so, the work transcends "greenness" and brings art into the realm of the "white," the ecological. Lorna Jordan may well be the first artist to conceive of Eco-Art as a form of infrastructure. Conceived as such, art becomes a foundational urban substance as essential to civic life as the water we drink and the air we breathe. It becomes, in essence, an integrative element applied seamlessly across the urban landscape from the scale of the region to the lone urinal, inverted or not.

The eco-works mentioned above were led by each artist. *Theater of Regeneration* involved the participation of an architect, a landscape architect, and engineers and other consultants. The initial request for proposals specifically

Figure 7.7. Theater of Regeneration *proposed "layers," "loops," and "lenses" as a form of infrastructure revealing Broward County's unique confluence of ecology and culture.*

ON PUBLIC ART

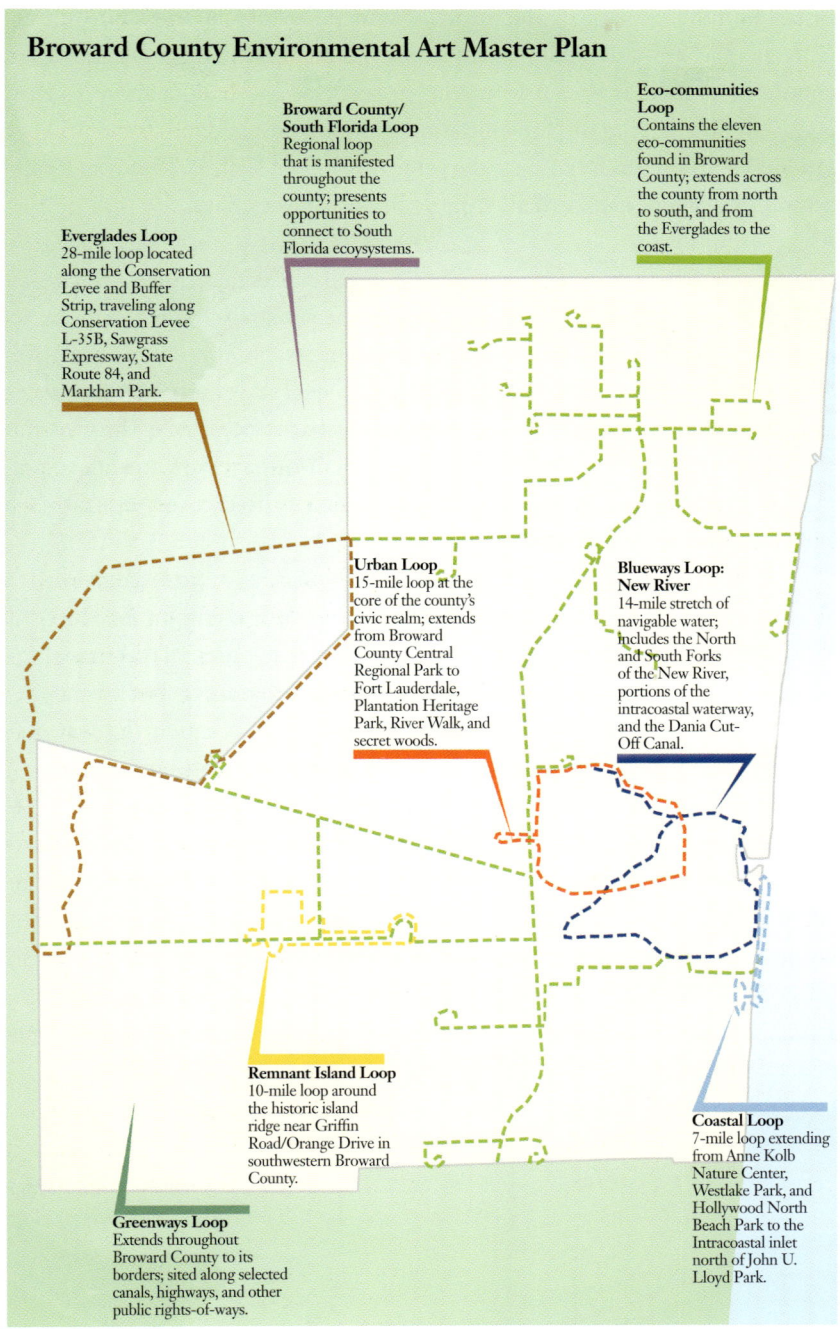

restricted the field to artist-led teams. Planners, architects, and landscape architects may well feel undervalued by such a mandate. The result, however, cannot be argued: it is transformative in scope as much as or more so than an urban plan relying on transportation, land use, or parkland as the primary guiding instruments. Regarding the value of art plans such as *Theater of Regeneration*, city planner and urban designer Todd Bressi explains:

> At one level, this new type of plan has helped determine which public-art projects are likely to have the most resonance, which will make the best use of available resources, and (in some cases) which are most likely to avoid controversy. But at their best, such plans may also serve as de facto urban (or civic) design frameworks. And in this capacity they have demonstrated a capacity to translate public-art programs to the scale of the city (or regional infrastructure like an airport, highway or transit line), shifting from incidental "percent-for-art" approaches to comprehensive strategies for building the public realm and addressing public narratives dispersed through time and space. In the process, they have drawn on knowledge of ecology, cultural geography, and social history to transform the way we look at cities — much the way City Beautiful plans imagined a new urban order a century ago.[14]

Public art must be integrated comprehensively in the city-making enterprise—especially as a *site-specific, utility-minded, and ecologically-based* development catalyst. Who leads and who follows is less important than obtaining the collaboration of curious and open-minded individuals who in the context of city design can spur and nurture the co-evolution of thought. Under such practice success would not be recognized simply as a matter of art, but rather as a matter of ecology, of constructing an urban nature.

Chapter 8. Dallas:
In Search of an Urban Future

Few regions exemplify the antithesis to a green infrastructure/localism/ public art-based urban ecology better than north Texas. Urban sprawl stretches unabated for tens of miles within multiple counties, with single-family communities and the strip malls hardly distinguished from one another, most all of it hanging off the backs of federal and state highways.

However, over the past few decades few other regions have sought to chart a different, more sustainable course. Commencing service in 1996, the Dallas Light Rail Transit system (DART) now has four lines totaling more than 85 miles of track connecting downtown Dallas with outlying suburbs, the largest such system in the country. Slowly but surely, the system is helping guide new development toward densifying transit nodes. In 2010, a regional plan promoting sustainability as an end goal was published by the North Texas Council of Governments (NTCOG) in cooperation with the Urban land Institute and the University of Texas-Arlington. *Vision North Texas* represents the culmination of three years' worth of consultation with community groups, professional associations and private stakeholders. The NTCOG also has adopted an integrated stormwater management system (iSWM) to help improve water quality, specifically through green infrastructure measures at the local level.

In Dallas, the Trinity River Corridor Project (TRCP) is the centerpiece of the effort to advance a sustainable agenda. The initiative encompasses more than a square mile of potential redevelopment land in the heart of the city, anchored by the transformation of a seven-mile-long section of the

Trinity River floodway into an urban park combining recreation, ecological restoration, flood control, transportation, and public art improvements (see figure 8.1).

The vision to transform the Trinity River Corridor into a green redevelopment catalyst did not materialize overnight. It came to light in 1992 through *Dallas Visions for Community*[1] in celebration of the city's sesquicentennial. Funded by the Dallas Foundation, the effort was led by local architect James Pratt in collaboration with the Dallas Institute for the Humanities and Culture. The Institute's director, Gail Thomas, contributed an impassioned foreword to the vision document:

> Why do we need to imagine parks and green spaces in a city? I would say the soul of the city needs the park so that its people will dream and play, and be silly and foolish, and be tender and stroll and kiss, and create fantasy worlds out of the clouds. When the problems, mistakes, broken promises, missed schedules, and caustic insults of the day get to be too much, I can imagine going to Trinity Park or Dream Lake, lying on the ground and remembering the poem by the poet Antonio Machado:
>
> > I dream last night, O Marvelous Error,
> > that honey bees were in my heart
> > Making honey out of my old failures![2]

Two years in the making and involving the participation of prominent local architects, planners, developers, and civic leaders, *Dallas Vision for Community* sought to reimagine the city as a greener, more humane environment. In so doing, it sought to redress the land use and development failures of old.

Planning Background

It is not for lack of planning that Dallas conjures images of boundless urban sprawl, tangled highways, smog-laden skies, and a downtown that largely empties after working hours. Soon after the great flood of 1908, city boosters retained landscape architect George Kessler to prepare a citywide

Dallas: In Search of an Urban Future

Figure 8.1. The Trinity River corridor cuts a seven-mile long, half-mile–wide swath through the heart of Dallas. The project constitutes the largest public work in the city's history.

urban plan. To guard against future flooding, Kessler proposed to taper and encase the Trinity River with 30-foot tall levees. He also proposed turning the envisioned floodway into a park. The levees were built over the ensuing two decades; the park was not, however, sealing the fate of the newly structured floodway as an inaccessible utility corridor. Kessler also envisioned a system of parkways associated with the Trinity River tributaries, but only one of these, Turtle Creek Boulevard, was implemented, the others succumbing to shifting political and development priorities.

Harlan Bartholomew followed Kessler with a comprehensive urban plan for the city in the mid-1940s. The plan addressed housing deficiencies, neighborhood boundaries, recreation demand, and water supply needs. It also recognized the rise of automobile travel by reaffirming Kessler's parkway scheme; but that, too, failed to generate the political will to build them. At the time, virtually no growth was envisioned past Loop 12, an at-grade beltway six miles out from the downtown. Today, growth in Dallas extends 30 miles northward past Carrolton and Plano to Frisco and McKinney.

Current suburban land is served by two newer beltways: Interstate 635, named after Lyndon Johnson, and the George H.W. Bush Turnpike. Connecting them to the downtown are two major spokes: the North Dallas Tollway and the Central Expressway. The latter was first proposed by Kessler and reinforced by Bartholomew, albeit as a green parkway in the mold of the Merritt Parkway in New York State rather than the 12- to 16-lane walled concrete behemoth that exists today.

Despite their best efforts, neither Kessler nor Bartholomew was able to keep Dallas from following the post-World War II urban growth model. In 1950 the city's population was 430,000; in 2010 it had reached close to 1.3 million. Sixty or so years ago it would have been unimaginable to conceive of Dallas as anything but an automobile-based metropolis affording a single-family house with a yard for nearly everyone. A fast forward video simulation of this growth would look like ripples in a pond spreading out from the downtown core in roughly 5.5-mile, 20-year increments: Loop 12 as the first (1950s); I-635 as the second (1970s); and the George H.W. Bush Turnpike as the third (1990s and 2000s).

In Horse Country, a Magic Hoof

It is against the backdrop of urban sprawl that *Dallas Vision for Community* emerges as a farsighted proposal; it bounces the radiating pattern of growth back to the first ripple of development, focusing on the core area of the city as the promise for a healthier, more humane environment. To forge consensus, the document set a 50-year planning horizon, enough lead time to quell the anxiety of established near-term development and political interests. It was also sold on the promise of water—not the potable kind, but that which is ensconced in the recesses of the western mind, namely the legend of Pegasus.

In Greek mythology the winged horse is said to have struck a rock in Mount Helicon with such force as to release the muses in a gush of water, creating a bountiful spring. Pegasus is thus associated with wellsprings as a source of beauty and poetic inspiration. Using the myth of Pegasus as a catalyst for downtown revitalization in the mid 1990s, Thomas spearheaded the construction of a half-acre plaza along Main Street bearing the name of the winged horse. Designed by artist Brad Goldberg, Pega-

Figure 8.2. Occupying a central location along Main Street in Downtown Dallas, the plaza is frequently used for concerts, rallies, and festivities.

sus Plaza has a cloud of mist emanating from slabs of indigenous Lueder's limestone as a centerpiece. Scattered about are stone engravings of each of the nine muse's particular vocation: dance, music, love, poetry, and so on (see figure 8.2). (Originally, the fountain tapped into an underground spring, but, citing water quality concerns, the city later turned to municipal water as a source.)

Framing the plaza to the south stands the Magnolia Petroleum Building, former headquarters of Mobil Oil. At the time of its dedication in 1922, the building was the tallest structure west of the Mississippi River. Today it houses a boutique hotel. But on its roof, attached to a once rotating scaffold, still stands the original symbol of Mobil Oil, a large, bright red figure of Pegasus (see figure 8.3). Its association with the igniting of power stems from the other trait for which the mythical horse is known: purveyor to Zeus of fiery thunderbolts. What a bold stroke, then, for Thomas to retake from oil the meaning of Pegasus—from automobile-teeming highways to a misting plaza and beyond to the promise of a civic wellspring—a "dream lake" on the Trinity River floodway. The connection is all the more apt in consideration of the horse's provenance: the blood of the Medusa, the snaked-headed Gorgon slain by Perseus, an all too easy allusion to the tangle of highways that encircles downtown Dal-

Figure 8.3. Pegasus Plaza helped rebrand the meaning of the winged horse and give new luster to its perch atop the fomer Mobil Oil building (today a boutique hotel).

las and has kept urban development separated from the city's birthplace, the Trinity River.

Breaking through the highway morass and retaking the Trinity River floodway as a lake-laden public amenity easily stands out as the most far-reaching proposal of *Dallas Visions of Community*:

> At the center of the river corridor, along the river's edge, the powerful skyline image of our downtown is our version of San Fran-

DALLAS: IN SEARCH OF AN URBAN FUTURE

cisco Bay or Rio Harbor. Trinity Lake would transform the present mud flats between the levees into a broad expanse of water at a scale to invite recreation and shore development.³

It was recognized that sizable improvements would be needed to realize such "shore development," including damming the river to create a sizable water body and bridging with development the freeway that stands between the downtown and the future lake. Given Dallas' previous failures to implement long-range plans, the outcome of this vision may have bordered on the utopian. But the idea of a more humane city with a lake-studded central park had wind on its back: the voices of admired city planners, architects, artists, teachers, and philosophers who for a decade prior had come to Dallas to speak at conferences on city planning at the behest of Thomas. Christopher Alexander, Christian Norberg Schulz, William H. White, Jane Jacobs, Robert Venturi, Charles Moore, Dan Kiley, and Yi FuTuan were among the leading voices. So, too, was that of Ivan Illich, who participated in two conferences. The first, held in 1984 and entitled "Waters and Dreams," inspired Illich to write *H₂O and the Waters of Forgetfulness*. In this work he emphatically separates H₂0, the stuff of science and utility, from the historical substance, the stuff of dreams and archetype:

> Following dream waters upstream the historian will learn to distinguish the vast register of their voices . . . He will recognize that the H₂O which gurgles through Dallas plumbing is not water, but a stuff which industrial society creates. He will realize that the twentieth century has transmogrified water into a fluid with which archetypal water cannot be mixed. With enough money and broad powers to condemn and evict, a group of architects could well create out of this sewage a liquid monument that would meet their own aesthetic standards. But since archetypal waters are as antagonistic to this new "stuff" as they are to oil, I fear that contact with such liquid monumentality might make the souls of Dallas children impermeable to the waters of dreams.⁴

Illich's words were a cautionary call to avoid simply damming the floodway as a means to put a reflective sheen on the city skyline. Rather,

Illich pleaded for Dallasites to "explore the moral and psychological consequences that will flow from the public display of recirculated toilet flush."[5] Illich was prescient, for in the end the city did indeed opt to use water from the treatment plant to feed the planned creation of lakes on the Trinity. He was imploring, essentially, for the city to produce "stuff" with poetic resonance.

Making it Happen

It would take six years after the publication of *Dallas Visions for Community* in 1992 for the planning and design of the Trinity River Corridor to begin. In 1998 Mayor Ron Kirk (who later became President Obama's U. S. Trade Representative) staked his leadership on the passage of a bond program to finance a *Master Implementation Plan for the Trinity River Corridor*, with parkland and a lake as the central elements. But oil, too, needed a seat at the table. Political and financial backing for the project required the accommodation of a toll road within the floodway—a couplet, with northbound and southbound lanes clinging to the downtown and Oak Cliff levees, respectively. Among the consultants engaged to produce the plan was Santiago Calatrava, retained specifically to design new bridges spanning the future park. (For the preparation of the Master Implementation Plan, Calatrava, along with WRT, was a subconsultant to Dallas-based Halff Associates.

(One evening, virtually alone in a barren school parking lot waiting to be transported to dinner following a public presentation, Calatrava proceeded to tell me of his design for the Kuwaiti pavilion for the World Expo in Seville. The building's roof was kinetic, opening and closing like the wings of a butterfly. Adaptation of animal physiology, he explained, was a guiding inspiration in many of his designs. Then a butterfly appeared out of nowhere and landed on his jacket lapel, gently flexing its wings as if to illustrate the point—positive evidence, no doubt, of the man's privileged standing with the muses.)

However, in the zeal to lay out the parkway, create new iconic bridges, and a large lake, the river itself was forgotten. *The Master Implementation Plan* proposed splitting the river into two channels paralleling the parkway couplets. So encased, the river was destined to further dissipate from the public imagination, as Illich had feared.

Such a scenario was antithetical to Thomas' tireless quest to humanize Dallas, to create a place of the soul, for the soul. She saw the river as the "water of dreams" every bit as much as a new lake. In 2003 her concerns took root with a new mayor, Laura Miller. Private funds were raised and new consultants brought to town to reexamine the future of the corridor. Led by Chan Krieger & Associates and Hargreaves Associates, a revamped plan emerged a year later after much public input and debate. The same elements as before remained: a body of water, only divided between two lakes connected by a narrow passage, an isthmus; the toll road, no longer a couplet but a now a single parkway just off the downtown levee; the river, not split but as a single channel with meanders, riparian edges, and adjoining wetlands; and parkland, only more of it and better, integrated with the lakes and river. Owing to the equipoise among flood control, transportation, environment, and recreation, the plan was officially labeled the *Balanced Vision Plan*.

In 2006 a team of consultants was retained to refine the new plan. (WRT was again retained for the work, this time as part of a team of consultants led by CH2M Hill, a national engineering firm.) More specifically, the effort aimed to finalize the corridor's recreation program, establish its physical character through comprehensive design guidelines (including the parkway), vet technical consideration sufficiently to establish engineering and cost parameters, and submit 35 percent design documents to the U. S. Army Corps of Engineers for review and approval. Special focus was given to green infrastructure and the integration of public art. Summarized below are the key design proposals resulting from this effort.

The Parkway

Central Dallas hosts a confluence of interstate highways, all vying for space as they circle the downtown before spinning outward to other regional destinations. Interstates 30 and 35E barrel towards this confluence over the Trinity River Corridor from the west, coalescing past the floodway in a maelstrom of ramps and lanes commonly referred to as the "Canyon/Mix Master," a woeful arrangement that induces chronic traffic back-ups. Alleviating the congestion was the principal argument in favor of the parkway.

To many residents, however, the presence of a high-volume, high-speed toll road running inside the city's long-envisioned park was anathema. "Do you want a highway in the middle of the park?" was the rallying cry behind a November 2007 ballot measure calling for the elimination of the parkway as a component of the project. It was narrowly defeated, but the vote strengthened the city's resolve to make the roadway as welcoming, unobtrusive, and green as possible. In pursuit of this goal, a host of green infrastructure measures were incorporated into the parkway design guidelines (see figure 8.4).

- 28 wind turbines earmarked for the lighting of the roadway and portions of the park;
- a 1.5-mile stretch of photovoltaic panels atop a required flood protection wall;
- 40,000 square feet of vegetated walls to attenuate sound and absorb vehicular emissions;
- five acres of water harvesting and biofiltration wetlands; and
- native vegetation along the roadway median and shoulders

Many highways are "architecturalized" by means of patterned formliners, colored surfaces and recurrent graphic motifs as a way to mitigate their hulking size and serial monotony. The LBJ Freeway (IH 635) and

Figure 8.4. Wind turbines, green walls, and stormwater filtration wetlands are among the green infrastructure measures embedded in the design guidelines for the Trinity Parkway.

Central Expressway (US 75) interchange in Dallas surely stand nationally as among the most prominent examples of this practice. By contrast, the Trinity Parkway will be dressed in green, a measure that will render it not merely context-sensitive, but context-setting.

Environment

Approximately 80 percent of the proposed parkland is earmarked for environmental restoration in the form of lakes, wetlands, riparian river terraces, meadows, bottomland forest, and urban reforestation. The table below summarizes the corridor's future land use.

Land Use	Acres
River	226
Lakes	301
Meadow	780
Wetland	317
Parkland	350
Playfields	119
Promenade	21
Parkway	114
Levee Park	86
Total	**2,314 acres**

Table 8.1. Trinity Parkway future land use

Through a system of meanders responsive to the location of lakes, wetlands, and park amenities, the river is poised to gain three-quarters of a mile in length, a far cry from its pre-Kessler natural form, but still a significant improvement. The meanders will create a more intricate and diverse landscape, supported by riparian edges that will enhance its function as a wildlife corridor from the confluence with the Elm Fork downstream to the Great Trinity Forest. This habitat will be terraced throughout the channel to simulate natural river morphology. The terraces will also graduate the river ecology in accordance with seasonal water level fluctuations.

The plan refinements also envision more than 300 acres of wetland, consisting of bioretention and filtration ponds, cypress bogs, lake-related marshlands, and extant seasonal ponding—a landscape that will enhance the floodway as waterfowl habitat for migratory birds. Some of the wetland is required as mitigation for the construction of the parkway and expanded sump areas outside the levees. All of it will provide the biological scrubbing of floodwaters.

Completing the environmental matrix are proposed meadow and associated bottomland vegetation. Native grasses and wildflowers will enhance the floodway's habitat for birds, butterflies, and other wildlife, while also serving as a colorful springtime canvas for hikers, joggers, cyclists, and horseback riders. More than 19,000 bottomland trees are proposed as part of this landscape, with Texas ash, buckeye, box elder, and pecan as dominant species.

Combined, the above habitats will transform the Trinity River floodway into a genuine patch of Texas blackland prairie, one that affords recreation in the context of natural habitats as well as infrastructural function from flood control, water quality enhancement, and carbon sequestration standpoints.

Recreation

As an urban park, the Trinity River Corridor has been designed with tens of miles of trails, playfields, play and picnic grounds, a park road for pleasure driving, and places large and small in which to gather and enjoy the green expanse. A major complex of sports fields is included opposite the western neighborhoods bordering the floodway, contributing a much-needed venue for active play and organized athletic competition. The new meandering river will provide future canoeists and kayakers a varied recreational experience. Proposed boat ramps will facilitate usage of the river, affording a 12-mile run from the northern boundary of the floodway (the confluence of the Elm and West Forks) downstream to Trinity Audubon Center in the heart of the Great Trinity Forest.

But the principal features of the recreation program are the isthmus-connected lakes, one labeled "urban" and the other "natural" (see figure 8.5). Both are fed by treated and recycled effluent, voiding the need for an environmentally disruptive dam. As implied by their names, the lakes are intended to fulfill markedly different recreational and environmental pur-

DALLAS: IN SEARCH OF AN URBAN FUTURE

Figure 8.5. Backed by new development spanning the levee and parkway, the urban lake is envisoned as the park's main recreational attraction. A cantilevered overlook, designed by Ten Arquitectos, would afford wide views of the lake and two Calatrava-designed bridges (seen here is an early proposal for the Interstate 30E bridge; a modified design is currently under construction).

poses: major public events and intensive recreation in and around the urban lake; more passive recreation amid a naturalistic setting in and around the natural lake. Among the features of the lakes are:

- Water-aerating "solar bees" and jets;
- A block-long overlook spanning the parkway and downtown levee;
- A mile-long promenade running the length of the urban lake;
- A "central island" between the river and the lakes functioning as an informal gathering and picnic ground;
- An amphitheater; and
- Dozens of ringed floating wetlands in the natural lake serving to biologically polish the lake waters.

PUBLIC ART

Artist Brad Goldberg was part of the design effort from the outset, working integrally with the project landscape architects, engineers, and other consultants in all aspects of the design. He was also entrusted with the development of a comprehensive public art program. It consists of three distinct kinds of art works:

- Ephemeral works placed temporarily in different areas of the park according to seasonal events and festivities;
- Council circles built out of native stone and placed in strategic locations throughout the park in celebration of the Native American tradition of communal life, learning and social exchange; and
- Iconic works by renowned artists at main gathering areas, such as the overlook.

However, Goldberg's singular contribution to the design of the park was the conception of the lake's isthmus. As a master sculptor with deep knowledge of the local bedrock, Goldberg shaped the isthmus as a series of limestone terraces through which course narrow canoe channels. Carved out of the limestone and shaped like idealized river meanders, the channels memorialize the lost (but to be partially recovered) river morphology. Here children and adults will step into the past, wading in new water that recalls the old (see figure 8.6). If Ivan Illich had lived long enough, he would surely have applauded such transcendent use of "toilet" water. More importantly, the isthmus is near the site where fordable limestone shoals along a river bend prompted John Neely Bryan to establish the town of Dallas in 1841. As recounted by historian John William Rogers:

Figure 8.6. Goldberg designed meandering channels carved out of native limestone as an isthmus between the urban and natural lakes, giving material form to the city's founding lore.

Dallas: In Search of an Urban Future

Bryan led his little party along the course of the river for some time until his eye was attracted by a diminutive bluff overhanging the stream on the east bank. It was a bare fifteen feet above water, but on the flat rolling country it was commanding. The crossing here was narrow and not too deep and the bottom land across the stream on the west bank suggested it was firm enough to bear traffic most of the year. The man climbed up the bluff and paused on top to look about him. At his feet were the river he had dreamed of and a natural crossing that Indians and other travelers would seek. Fertile prairie broken by gentle woodland met his gaze on all sides. Here was the answer to that urge which had brought him so far.[6]

The site of Bryan's "bluff" lies close to the foot of present-day Commerce Street, just south of the "tripple underpass" of lore. His cabin was built above the river bank near present-day Dealey Plaza, a place that is more commonly known as the "grassy knoll." The proposed isthmus, then, signifies not just the memory of the landscape that was, but also the profound and indelible relation between nature, infrastructure, and culture, and, arising from it, the events that change the land, indeed the world.

Economic Development

In the context of urban development, few things can crystallize the potential for change better than proximity to parkland and water. The Trinity River Corridor Project thus represents a massive "game-changer," one that will forever alter the urban character of central Dallas. Facilitating the coming transformation is nearly a square mile of redevelopment land at either side of the floodway, representing approximately 40 million gross square feet of new potential development (assuming 50 percent coverage and a floor area ratio of three). Such scale of development represents an investment of about $7 to $8 billion, a six-fold value of the estimated cost of the park.

Construction approval for the project is expected in 2014, yet the development community has already begun to capitalize on the promise of a transformed floodway. Among various development proposals are a 3.84

million square-foot mixed use project adjoining the park's future overlook, anchored by an iconic 55-story tower (see figure 8.7); a new 60-acre midrise sustainable community across the levee from the natural lake and upholding the NTCOG iSWM principles for stormwater management; and infill development within 100 acres of marginal industrial land in West Dallas at the foot of the Margaret Hunt Hill Bridge (Calatrava's only completed Trinity-related work thus far). This last development is highlighted by food incubator stalls and marketplace occupying a former industrial shed.

Distributed across the redevelopment area, 50,000 new residents could well be living in close proximity to the downtown in the coming decades. Coupled with planned development within the downtown itself and reinforced by new parks in its midst, the core area of the city could well absorb one-third of anticipated growth over the next 20 to 30 years, more than fulfilling the promise of *Dallas Visions for Community*.[7] Moreover, per city initiatives related to public transportation and energy efficiency, this growth is destined to be LEED-certified, and transit- and pedestrian-oriented. New residents will create the demand for markets, schools, day care, and retail, and entertainment facilities to the benefit of adjoining communities, especially in impoverished West and South Dallas. People will be able to travel from one side of the city to the other through paths and trails, on foot or bicycle, easily up and over the levees and across a regional

Figure 8.7. In this montage, the Downtown skyline is reflected over a future "dream" lake. Proposed new development is shown on the far right, anchored by a 50-story mixed use tower.

recreational hub. The services and attractions on one side of the river will be reachable through the park to residents on the other.

Initiatives are also underway to extend the system of trails outward from the corridor to the heart of Oak Cliff along Coombs Creek; to East Dallas and beyond to White Rock Lake along the Santa Fe Trail; to the Design District along the Trinity Strand trail; and to the heart of West Dallas along Bernal Trail. George Kessler envisioned parkways as a defining and unifying urban landscape for Dallas. Greenways and trails are emerging as the twenty-first century equivalent. When completed, tens of thousands of residents will live within a few blocks of these greenways and, therefore, be within bikeable reach of the "dream lake."

And about public art? Although some new works have been installed at entry corridors to the park and others are under consideration at key gateways, the city has yet to embark on a comprehensive art plan for the entirety of the Trinity River Corridor redevelopment zone. But one thing is clear: once the park is developed as planned, every drop of city tap water will be in play as the "stuff" of art, cycling from drains, pipes, valves, and pumps through the city's treatment plant into the natural lake and over the isthmus to reach the longest river in Texas as the substance of myth and fantasy. Much of this tap will have been generated by new development, helping cast Dallas as a national model for the inevitable transformation of our cities into denser and eminently desirable urban environments.

Chapter 9. Toward a Climax City

Planning is a philosophy for organizing actions that enable people to visualize the future. It is up to us to plan with vision.
—Frederick Steiner, *Design for a Vulnerable Planet*, 2011

The term "climax" refers to the ecological condition of sustained balance, or "steady state" between a biotic community and the environment in which it exists. Climax is also the name of a village where 700 people live in farm country between Kalamazoo and Battle Creek, Michigan. In 1838 Hiram Moore invented the horse-pulled harvester combine there, a precursor to the present day wheat-harvesting machines that keep much of the nation's native grassland in perennial check. Perhaps in observance of the machine's mowing prowess the village's original name—Climax Prairie—was eventually foreshortened, burying in local lore the town's association with the cells of native tall-grass prairie that once thrived in Southwestern Michigan. Yet the town's name remains fitting: despite a few ups and downs, the village has not appreciably changed in more than a century, a fair representation of a steady state urban condition.

But what of a big city? Could large urban areas achieve a climax condition given the constant influx of people and goods and services, the ebb and flow of businesses and organizations, and the constant pull and push of global commerce and its impact on growth and decay? An ecological climax predicates the sustained adaptation of organisms to a given set of environmental conditions, climate being among the most critical. If it is accepted that the environment of an urban area—*a densifying urban area*—should be conducive to a sustained human adaption (presuming health and well-being the aim), then there should be an optimal way to build it—or *build toward*.

147

The notion of ecological climax can be challenged by scientific evidence to the effect that natural systems are in constant flux—that a climax community in nature is nothing more than a temporary state of biotic balance within an evolutionary arc.[1] Still, a climax state ultimately represents an efficient use of resources (sunlight, water, nutrients, etc.) to the sustained benefit of flora and fauna at a given point in time. The question is: how should cities, as human habitat, deliver an optimal environment at *this* point in time, given present needs and aspirations? As presented in this book, such an optimal urban state depends to a great extent on the fusion of green infrastructure, localism, and public art. The redevelopment of the core area of Dallas as catalyzed by the Trinity River Corridor Project offers a window into this world, long as it may be becoming a reality.

In considering how to progress towards a climax urban state, three operational scales become useful: that of the CITY, as an agglomeration of urban centers, districts, and neighborhoods; that of COMMUNITY, as the space within which the lives of discrete populations unfold on a daily basis; and that of BUILDING, as a distinct development program within definable site boundaries. Owing to the value of open space in advancing urban quality and identity, it is further useful to accord to these scales a corresponding planning and design typology, namely LANDSCAPE, PARK, and GARDEN, respectively. Diagrammatically, then:

A planning and design hierarchy is thus established, facilitating the consistent fusion of infrastructure, localism, and public across scales of development. Following are examples of existing and/or envisioned cities, communities, and buildings in support of this thesis.

Toward a Climax City

City as Landscape

"Landscape" is the representation of nature and culture as interdependent and co-evolving producers of what we see as we move about the land. Embedded in "landscape" is the beginning of everything—the raw land that became settled as a result of human imagination and toil. The ecology that underlies this land, whether extant or not, can become deeply held as part of the story of community.

Knowing this ecology was central to McHarg's pedagogy. In *Design with Nature*, for example, the physiography of Washington, D.C., is presented as a way to explain the relationship of federal facilities to the underlying bedrock. Of note is the U.S. Capitol. The building rests on a high point along the Pamlico-Wicomico geologic escarpment, a feature that fully supports the civic need for symbolic prominence.[2] In time the lowlands west of the escarpment became the National Mall, portions of which flood during heavy rainfall. The National Park Service has planned a floodgate near Constitution Avenue to prevent rising waters from extending past the Mall into the city. Near this gate is Constitution Gardens, a 50-acre site created for the national bicentennial in commemoration of the American Revolution. Its curvilinear pond and surrounding informal landscape embody a faint echo of the natural habitat that existed close to the Capitol and White House at the time L'Enfant planned the nation's capital. In 2012 the Trust for the National Mall chose new designs for Constitution Gardens through a national design competition. The winning entry, by Peter Walker & Partners in association with Rogers Marvel Architects, reinforces the informal quality of the landscape through curving paths and rolling topography. It also proposes a ring of emergent vegetation at the edge of the pond, a green infrastructure measure that recalls the marshlands that once defined the setting of the nation's capital.

Places like the National Mall that in parts evoke or reveal a city's founding landscape foreshorten the path between the eye and the soul. Doing so sustains the link between past and present, according urban places a measure of poetic resonance. In Fargo 365 the prairie potholes served to establish this vital poetic link. In Dallas it will be the recovery near the downtown of hundreds of acres of Texas Blackland Prairie as part of a reconstituted river corridor. Cities such as Orlando and San Francisco provide even broader examples—Orlando through parkland that surrounds many of the notorious limestone sinkholes that dot the city, with Lake Eola, the oldest and most perfectly shaped, being a downtown landmark, and San Francisco through a street grid that extends over steep hills without regard to contour or slope, magnifying the terrain over which it lies. In both cases, the geologic substrate has become a civic *cause celebre*.

San Diego provides another example of an urban identity molded by natural features. The city is characterized by finger canyons that cut through the gridded urban mesa; by broad valleys that collect major drainage, transportation, and big box retail; and by embayments that capture the downtown skyline and major recreational and visitor attractions. In the classic 1974 study, *Temporary Paradise?: A Look at the Special Landscape of the San Diego Region*,[3] Donald Appleyard and Kevin Lynch defined the geo-physical structure of the city, calling for the preservation and enhancement of the bays, valleys, and canyons as essential city-shaping elements. Few studies so clearly establish the power of, and dependency on, a city's foundational landscape as a guide towards growth and development.

Louisville

Temporary Paradise? owes its thesis to the legacy of Frederick Law Olmsted and the method by which he planned urban parks and open space systems. In 1891 Olmsted proposed a comprehensive array of parks in Louisville, Kentucky, the last such system he would conceive before retiring as a practicing landscape architect. He strategically located and purposefully designed the city's three main parks around specific aspects of the region's ecology: for Cherokee Park, the Bluegrass Savanna and stream valley; for Iroquois Park, the Kentucky Knobs or mountain region; and for Shawnee Park, the Ohio River bottomlands. About these sites Olmsted recorded specific attributes:

> In Cherokee Park is preserved a fine example of the bluegrass country which has made Kentucky famous. It includes massive, rounded, grassy slopes down to the narrow winding valley of Bear Grass Creek, something which the other parks do not possess . . . In Shawnee Park the main feature is its relation to the great Ohio River. In addition to this it has the landscape value of preserving

Toward a Climax City

Figure 9.1. Baringer Hill, one of many distinctive features in Cherokee Park designed by Olmsted, embodies Kentucky's bluegrass country.

the quiet, charming effect of the best type of bottom river land, a broad bluegrass meadow with open groves and scattered trees, an effect different from that obtainable in the other parks[4] (see figure 9.1).

Through these parks Olmsted laid anchor to the city over its dolomitic bedrock, in effect elevating the primordial landscape as a source of civic identity (at least as regarded by early European settlers). The Olmsted parks, in effect, represent the city's state of beginning as an ingrained abstraction, one that remains vivid, like a cherished work of art, by the region's immaculate horse-breeding landscape and, architecturally, by the famous hippodrome that annually showcases its finest offspring.

Presently, Olmsted's legacy in Louisville is being expanded upon by 21st Century Parks, a local nonprofit organization that has led the creation of new parkland on the eastern edge of the city. Called the Parklands of Floyd's Fork, the 3,800-acre initiative seeks to preserve both the natural and agricultural heritage of the region while providing recreation and trails along an 18-mile section of the fork's meandering valley.[5] The planning and design of the parklands has been led by WRT in association with Bravura, a Louisville architecture firm. Dan Jones, head of 21st Century Parks, intently envisions the Parklands as a future *central* open space, much as Olmsted saw the city growing around his parks more than a century ago. Through the Parklands and other open space initiatives, Louisville rightfully claims the title "City of Parks." More accurately, in its open space Louisville captures the most endearing qualities of the region's landscape, a rooted imprint that has and will continue to shape the way the city grows for generations.

New Orleans
The City of New Orleans offers an example of how a foundational landscape element—water in this case—can potentially reshape urban growth as green infrastructure. The city was first established over a natural rise along a bend of the Mississippi River, high ground that protected what today is the French Quarter from recurring floods. As the city grew and development extended into the lowlands, vast defenses were built to keep floodwaters at bay, such as levees, drainage canals, and stormwater pumping stations. In 2005 Hurricane Katrina tragically exposed the limitations of this system. Much of the area became quickly inundated by 4 to 12 feet of water. Hundreds of thousands people were displaced, leaving in their wake the prospect of a much diminished city. A commission of civic leaders was established to jump-start the reconstruction and reinvestment process immediately after the devastation. A scant four months after the hurricane, the Bring New Orleans Back Commission (BNOB) unveiled an $18 billion recovery plan. (The recovery plan, called *Action Plan for New Orleans, the New American City*, was led by WRT, with John Beckman in the capacity of lead planner; I contributed the plan's approach to parkland and open space.)

The plan was based on three interlaced strategies:

1. improved flood and stormwater protection;
2. improved transit and transportation; and
3. improved parks and open space.

The first strategy focused on the upgrade of perimeter levees, new pumps, and floodgates; the development of internal levees with separate pumping stations (like the sectioning of a ship's hull); and the restoration of coastal wetlands as a natural line of defense against storm surge, a key green infrastructure measure. The second strategy, transit and transportation, focused on the expansion of planned light-rail and new roadways with "neutral grounds," or medians, devoted to pedestrian trails and bikeways, another green infrastructure measure. Beyond accruing energy efficiency, the transportation strategy also addressed the need for more effective evacuation in the event of future disasters. Lastly, the open space and parks strategy called for new parks in every district as redevelopment catalysts, new greenways connecting neighborhoods and employment centers, the use of canal edges and canal covers as open space amenities, extension of canals for internal drainage and navigation purposes, and use of lowland as part of an improved stormwater management system. The latter measure identified the areas subject to extreme potential flooding as places suitable for groundwater recharge, which would help arrest the rate of land subsidence (see figure 9.2).

A subsequent planning exercise ensued, the Unified New Orleans Plan (UNOP). The purpose of this effort was to consolidate all ongoing reconstruction initiatives (including the BNOB Commission plans and the City Council's Neighborhood Planning Initiative) into a comprehensive set of recommendations. Many community workshops were conducted under the

Figure 9.2. This rendering represents a summary of the open space recommendations presented to the Bring New Orleans Back Commission. Included are parks, greenways, boulevards and canals as a comprehensive, hydrology-based green infrastructure system.

UNOP in the zeal to rebuild, and rebuild quickly. Underlying the effort was the political mandate to not prescribe where people could or could not reside, regardless of topography, fiscal impact, environmental constraint, or reconstruction timeline. This was understandable—land tenancy in New Orleans is deeply rooted, especially in lower income areas. No plan superseding generational property rights could have survived. The UNOP respected such tradition by promoting reconstruction practices that could withstand future potential flooding in accordance with FEMA guidelines, namely that homeowners could rebuild at will, provided that finished floor elevations were kept a minimum of three feet above existing grade. Implicit in this allowance was the belief that the city's improved levees would withstand a direct hit from a future Category Force 5 hurricane.[6]

Lost in the rebuilding shuffle, however, were the initial green infrastructure measures related to parks and open space. Only the BNOB Commission plan proposal for the use of "neutral grounds" as a community amenity is noted in the UNOP report. Had the New Orleans reconstruction effort been fully guided by green infrastructure, the city's post-Katrina urban form would have established a more tangible link between the body of the city and its soulful connection to the founding landscape—the former through a hydrology-responsive landscape of canals, parks, and greenways, the latter through the collective sense of resilience accruing from it. Rebirth, more than reconstruction, would have been the order of the day. Had public art been integrated with the planning process, water, in all its physical and fabled manifestations, would have plied across the urban landscape, rendering New Orleans a hydrologic "Theater of Regeneration" in the manner of Lorna Jordan's public art plan for Broward County.

Since Katrina, the city's population has gradually increased. But at 365,000 (as of 2012), it stands at about three-quarters of what it was before the hurricane struck. Few people have returned to the Ninth Ward, the area hardest hit by flooding. Fewer than 6,000 people live there today, about a quarter of the pre-Katrina count. Amid abandonment and desolation the prospects for new development in the Ninth Ward remain uncertain. Much of the area is likely to revert to permanent open space, whether planned or not. In the face of such lingering trauma, New Orleans as a whole may yet wait decades, if not centuries, to achieve a "climax" urban condition.

Community as Park

As a physical realm, "community" signifies the space where discrete populations move about daily to and from home, work, school, and services. Historically, such a realm has been largely treated as the antipode to urban parks—the gritty environment that makes green open space a necessary exception in the first place. The value of New York City's Central Park derives to a great extent from the greenery it offers in contrast to the built-up blocks that surround it. As stewards of the park, the Central Park Conservancy proudly promotes such distinction, referring to the park as "a green oasis in the great concrete, high-rise landscape of New York City."[7]

But what if the "grit" of the city were a green amenity in and of itself? What if it became park-like? Philadelphia intends to add 500 acres of recreational space within existing neighborhoods, as called for by its *Green 2015* parks master plan. This is commendable, yet in Philadelphia as in most cities, finding cost effective space for new parks within a safe and amenable walking or biking distance from most people's homes remains a challenge. Efforts such as the Georgetown Waterfront Park are uncommon, initiatives such as the Trinity River Corridor Project, exceptional. Therefore, future outdoor recreational needs must be met by also maximizing the recreational and green (i.e., "living") quality of the everyday environment—the sidewalks, trails, alleys, streets, avenues, boulevards, and odd leftover public areas that typically comprise approximately half the land in any given city. Owing to their potential as active mobility corridors, streets and greenways become especially important as green amenity venues.

"Living" Streets

Much can be said about the benefits of the national Complete Street program that is slowly but surely changing the character of the nation's urban thoroughfares. This initiative addresses modes of transportation, while its environmental cousin, Green Streets, is focused primarily on stormwater management. Under a "climax" scenario, streets would function as livable, everyday amenities, an extension of a city's mobility, environmental, and recreational matrix. "Living Streets" should invite an elderly person and her grandchild to sit and rest, to say hello to neighbors and friends, to smell a flower, hear birdsong or interact with artwork. Street art, from graf-

Figure 9.3. Facing small neighbrohood eateries, this parklet along 44th Street in West Philadephia has become a favored social hangout.

fiti works, murals, and temporary sidewalk paintings to word art, sculpted furnishings, and periodic art performance should by fully integrated with Complete and Green Street programs.

There are streets in West Philadelphia where children create chalk art on slate pavers, play four square on makeshift sidewalk line work, swing from street trees between parked cars, test their balance walking on narrow garden curbs, and upgrade their skateboard skills on warped sidewalk paving. Owing to the ubiquity of front porches, entry steps, and the occasional sidewalk bench, these streets also function as socializing foci for young parents and grandparents. The addition of seasonal curbside parklets outside many cafés and restaurants is creating further opportunity for social exchange (see figure 9.3).

Features such as rain gardens, bioswales and filtration wetlands should be a complement to everyday recreational walkability. They add greenery, to be sure, but they also help develop social capital through community-based maintenance programs. Traffic control measures that promote pedestrian safety must be equally complementary: smooth paving in accordance with the American Disability Act; consistent street parking as sidewalk buffer; corner curb extensions that shorten crosswalks (an especially useful measure for slow-walking older people); pedestrian refuge zones on multilane arterials; table-topped driveways that maintain even sidewalk grades; signalized midblock pedestrian crossings; and countdown intersection signals.

Design proposals for Raritan Avenue in the city of Highland Park, New Jersey, were designed following the "living" street model. The improvements addressed the public space component of a revitalization plan for the town's principal commercial corridor. Of special note are "living room" street corners with sofa-like benches, side tables, and art crafted carpets of patterned recycled glass, a symbol of the community's toddler-to-grandparent sociability and welcoming multiculturalism (see figure 9.4). The street corners flair out behind side-street rain gardens that collect localized

Figure 9.4. A "living room" bump-out at a Raritan Avenue street corner affords a social space, functioning also as a traffic calming device. A rain garden, to the right, captures roadway and sidewalk stormwater.

stormwater, an expression of this small town's outsize belief in a sustainable "steady state." (The Highland Park Streetscape was spearheaded by former Mayor Meryl Franks as part of a 2005 downtown revitalization plan. Both the streetscape and revitalization plan were prepared by WRT.) To be sure, not every street in every city needs such treatment. But every community should have such corridors, especially along high traffic pedestrian and bike areas associated with schools, services, shopping, and local parks. This is a growing need: in market preference surveys, walkability and "life cycle mix" are listed as the top desired conditions of a healthy neighborhood—by far.[8]

Greenways

Intercommunity greenways are a necessary complement to park-like living streets. A century ago, off-street pedestrian and bicycle corridors unrelated to parks did not exist as single use public rights-of-way. Today they occupy a special green infrastructural niche, one that also abets urban regeneration. The Katy Trail in Dallas is a prominent example. Running for 18 miles between the community of Highland Park and the Victory Park development near the city's downtown, it is pressed between residential, office, and commercial developments that were spurred in part by the trail's active mobility and recreational appeal. Only a single at grade roadway crosses it, a benefit to families with young children seeking a safe and casual bike ride. Close to Victory Park the trail descends along an exposed limestone bluff, a rare occurrence in Dallas outside river corridors and stream banks. In doing so, the Katy Trail provides Dallasites an Olmstedian window into the city's foundational landscape, a place for the spirit to rise as the body speeds by on a downward glide. In time, the Katy Trail will be extended to the Trinity River Corridor and beyond to the planned 1,668-mile, 166-city, 10-county North Texas veloweb.[9] This system will connect many of the region's major parks and natural preserves, establishing a physical link between the CITY and COMMUNITY scales of city building.

Blended with green infrastructure, greenways can also become conduits for art and culture. This was part of the impetus behind the creation of the Indianapolis Cultural Trail[10], a legacy of Gene and Marilyn Glick. (The project is the brainchild of Bryan Payne, president and CEO of the

Toward a Climax City

Figure 9.5. M12, a collective of artists and designers, created "Prairie Modules 1 & 2." The work invites people to reflect about the juxtaposition of technology with the Indiana agricultural heritage.

Central Indiana Community Foundation. Kevin Osburn of Rundell Ernstberger Associates, LLC, an Indianapolis planning and design firm, led the design of approximately $48 million in improvements. The Glicks contributed $15 million to the project.) The eight-mile trail encircles Indianapolis' downtown, connecting several neighborhoods with the outlying network of regional greenways. It consists largely of adjoining pedestrian and bicycle ways, in parts carved out of existing traffic and parking lanes.

Destinations along the trail include the state Capitol, convention center, public library, American Legion Hall, Indiana History Center, Eiteljorg Museum of American Indian and Western Culture, and Madame Walker Theater Center (a locus for the celebration of African American culture), plus scores of shops and eateries. Approximately three-quarters of an acre in rain gardens retain and filter stormwater throughout. As of 2012, more than a dozen works of art have been placed within the trail corridor. One of them displays poetry on trail bus stops; another casts welcoming shade via a trellis supporting photovoltaic panels that feed the city's power grid (see figure 9.5); and another emits floral scents from a pavement grate that references the city's historic coal vaults. Twelve sculptural gardens pay homage to 12 "luminaires" for their contributions to science and art, among them Mark Twain, Booker T. Washington, and the Wright Brothers. In the words of Mindy Taylor Ross, the project's public art coordinator, "the Trail sends a signal that [this] city values the ability of creative people to impact the quality of our public spaces and our lives." Under a "climax" urban scenario, urban mobility becomes an art-suffused experience.

Building as Garden

From the early paradise gardens of the Middle East, the Pompeian courtyard, the medieval cloister, the Moorish palaces of southern Spain, the English winter garden, and the modernism of Wright and Alto, to present day "green" architecture, the need to bring a measure of healthy air, light, water, and greenery into the built environment—to inhabit, in effect, a garden—has been a recurrent building tradition.

Few architects have explored the promise of the garden archetype in buildings more keenly than Carlo Scarpa. Architectural historian William Curtis regards Scarpa's work as touching "a timeless score . . . emerging from the depth of the mind, giving shape to myths that have a universal dimension."[11] Of all Scarpa's works, his Villa Ottolenghi is perhaps closest to Curtis' characterization. The villa (in reality a normal home), is set in the hillsides of Lake Garda in Lombardy, not far from the town bearing the lake's name. From the roadway, an innocuous sunbaked rooftop terrace serves as foreground to the cypress-laden rural landscape that surrounds it. Past the entry gate a stepped fissure in the terrace is revealed, leading visitors downwards into deep shade, past a trickling and humidifying waterfall, and around a bend with cascading greenery to the main door. The passage from dry heat to comforting coolness is as measured as it is absolute.

Inside, the embrace between building and landscape comes into full view. A small interior pool captures the reflection of the garden through a picture window; it also extends into the garden past the dividing glass to form a two-tiered body of water. Thus, body and mind are bridged—outside, the quintessential place for sensual immersion and relaxation; inside, the withdrawing nook for creative reflection. The pool, in essence, materializes the upper and lower floors of the Leibnizian house, a testament of the "garden" as both a real and ideal ecology. A water runnel through the exterior wall connects both tiers of the pool, underscoring the indivisibility between the interior ("upper") and exterior ("lower") realms of the ecological house. Scarpa achieves the same effect in the renovation of the Palazzo (and garden) Querini-Stampalia in Venice – possibly his most refined work (see figure 9.6).

Arguably, the green architecture movement began in Malaysia in 1993 with the completion of Menara Mesiniaga. Designed by Kenneth Yeang, the 12-story building employs natural ventilation techniques and the multistory integration of vegetation to effect cooling.[12] Today, the integration of infrastructural greenery in buildings is becoming ever more prevalent, especially for institutional commissions. The U.S. Coast Guard Headquarters complex in Washington, D.C., for example, sprawls across the hillsides of Anacostia like a modern-day Alhambra, with water-filtering green roofs and water recycling ponds diffusing the line between building and landscape. In San Francisco, the California Academy of Science is topped by an expansive, swelling vegetated roof as if the surrounding coastal range had washed over it. And in Washington, D.C., the Sidwell Friends Middle School is hailed for the use of its gardens as part of its mechanical apparatus. Here, a series of constructed wetlands absorb, cleanse and filter building effluent and exterior storm runoff, producing gray water for use in toilets and the school's cooling tower.

But "green" in and of itself does not constitute a "climax" architectural state. Artistic intent must be part and parcel of the garden ideal. As architect, artist, and educator James Wines argues:

> Too often the problem with so-called green architecture is the conflict between having an admirable commitment to ecological design ... and a failure to convert [such] noble objectives into an equivalent artistic expression.[13]

In this regard, Wines's work is both singular and pioneering. As early as 1981 his firm, SITE, proposed "Highrise of Homes." The conceptual proposal consists of a 12-story building for "single-family homes," one or more per structural bay, landscaped yards included. As a "matrix of housing choices" the work represents an antidote to the alienating, impersonalizing

Figure 9.6. At the Fondazione Querini Stampalia, a fountain spans the transition from the garden to the building interior. With integrated planting, the garden portion of the fountain embodies nature (the body), while the interior elevates it to the realm of the abstract (the mind). Both floors of the ecological house are thus expressed.

Toward a Climax City

Figure 9.7. James Wines's museum proposal integrates two existing mosques, one at the center of the complex, creating, in effect, a building within a building.

effect of industrial architecture. (Wines's *Green Architecture* is among the first comprehensive reviews of the genre.) Realized examples of Wines' fusion method were made manifest through the design of several BEST Products stores. In Richmond, Virginia, one of them evokes nature's revenge on humanity's degrading consumerism by simulating an architectural ruin overtaken by wild vegetation.

But it is Wines's proposal for the Doha Museum of Islamic Arts where the fusion of building and landscape achieves greatest clarity. Here, vertical roof undulations gradually morph into horizontally undulating garden plots. Such play recalls a Mobius Strip, one of horizontal surface becoming seamlessly vertical, achieving a smooth transition between a site's interior and exterior realms. Furthermore, narrow slits of glass separate distinct bands of the undulating roof, evoking the shape of the new moon, archetypal symbol of Islam. This last feature accords the design incontrovertible rootedness, a sense of poetic localism that exalts the human spirit at the highest level of existence (see figure 9.7).

Figure 9.8. Individualized dwellings and integrated greenery gives the Hundertwasser House a sense of environmental and artistic fluidity—a par excellence expression of the "building-as garden" notion.

To be sure, the building-as-garden component of a climax city is not restricted to single-family homes or major institutional works; it can also exist as multifamily housing. In a Vienna apartment building bearing his name, artist/architect Friedensreich Hundertwasser showcased the fusion of art, greenery, and architecture as never before conceived or attempted (see figure 9.8). The result recalls the Baroque in its fluid and exuberant plasticity. Soon after completion a "tree-tenant" emerged from an upper

Toward a Climax City

story window. Reaching out tenuously from the interior darkness, the tree seemed to beg a reassessment of humanity's building ethics, at once blaring the call for a healthier way of living and reminding us of our paradisiacal (but lost) beginnings.

If matters of the spirit inspire art, and if art anticipates the world as it could be, then Hundertwasser should occupy a privileged seat in the pantheon of city-building prophets. His art production welled from a deep concern for the five dimensions of the human skin—the body each one of us inhabits, the clothes that give it comfort, the home that protects it, the society of which it is a part, and the planet that sustains it. He expressed this concern through art—at first two-dimensionally in painting and later three-dimensionally in habitable buildings. Hundertwasser reserved praise for Gaudi and nouveau architecture, styles that in their playful fluidity capture the idiosyncratic bonds we share as individuals and, collectively, with the planet as our ultimate garden. Damnation was directed towards the international style and the industrial, straight-line rigidity by which it forced people to live sullen lives. In 1956 Hundertwasser published the "Mouldness Manifesto," stating that:

> The tangible and material uninhabitability of slums is preferable to the moral uninhabitability of utilitarian, functional architecture. In the so-called slums only the human body can be oppressed, but in our modern functional architecture, allegedly constructed for the human being, man's soul is perishing, oppressed. We should instead adopt as the starting point for improvement the slum principle; that is, wildly luxuriantly growing architecture, not functional architecture.[14]

Hundertwasser's words are as cautionary as those of Ivan Illich with respect to Dallas's "Dream Lake"—mere functionality is not enough. Green buildings must embrace the human spirit as a means of transcending physical well-being. Like gardens, green buildings must exhibit a fractal array of solids and voids, each loop tracing the larger body of infrastructure, and art as a unified and sustaining habitat. The Fargo 365 Urban Block competition is emblematic of this ideal. The design melds public and public space, from street side to building rooftops, forming a climate-mitigating, art-laden, and green fabric that serves as a poetic and utilitarian reminder of the ecology that resides in the collective mind. Implied by this design is the garden archetype—a place that comforts the body, lifts the spirit and provides space for social exchange. Under a building-as-garden practice, no two buildings would be alike, each an efflorescence of localized bedrock, both concrete and abstract. In this context, planning and design become exercises of discovery and invention, of marking ground as a possibility of green and artful composition. Such would be life in a climax city: an existence at the seam of the house divided, from the immensity of clouds to the immediacy of a leaf.

Chapter 10. Beyond, Ahead

> We have to change our priorities in the spending of our individual and national wealth. If we want to stop pollution, sprawl, destruction of land, then we, individually and collectively, must be willing to pay for it. There is no magic formula, no miracle to technology, no wonder material, no automation, or any other trick to sidestep this fact.
> —Moshe Safdie, *For Everyone a Garden*, 1974

Design with Nature represents a milestone in the era of emerging environmentalism. At a time when the national drive towards suburbia was gaining acceleration, McHarg's book caused people to pause and ask "how can the trail be blazed without searing the landscape?" His signature achievement, the ecological planning method, provided an answer. Since 1969, tens of millions of homes have been built in the countryside, remaking the rural landscape into a sprawling haven for single-family homes, sensitive ecological resources duly preserved.

Today, the triple bottom line sustainability objectives of environmental health, economic development, and social equity call into question the viability of suburban development as a societal aim. The fuel consumed in personal transportation, the ever-longer commutes, and the inefficient dispensation of municipal services required to sustain far-flung communities is no longer economically or environmentally viable. Regardless of its correctness from an ecological standpoint, no amount of spared woodland, wetland, erodible slopes, riparian stream banks, or aquifer recharge zones can mitigate the health and social impacts of distant subdivisions that induce automobile dependence. The counter-benefits of a half-century of sprawl, coupled with the unrestrained consumption of energy in buildings and transportation, have placed the nation at a tipping point. As Thomas L. Friedman noted in a *New York Times* editorial:

> With Europe in peril, China and America wobbling, the Arab world in turmoil, energy prices spiraling and the climate chang-

ing, we are facing some real storms ahead. We need to weatherproof our American house—and fast—in order to ensure that America remains a rock of stability for the world.[1]

Weatherproofing the American house — the nation — requires a fundamental rethinking of the national urban landscape. Smart growth provides vital guidance—more compact, walkable, transit-oriented communities that preserve open land represent positive development goals. Equally important is denser placemaking for home and work that proves to be eminently more desirable than a suburban alternative.

The opportunity at hand matches the challenge: the Lincoln Institute of Land Policy reports that "the United States will require nearly three-quarters as much new built space over the next two decades—homes, offices, stores, warehouses, and so forth—as has been built over the past four centuries."[2] A vast new array of places can therefore be designed to meet the imperative for better, healthier, and more efficient cities, from the breakfast nook to the workplace and every mode of public infrastructure in between. In denser towns and cities the possibilities for ecological placemaking will multiply, expanding the opportunity for individual and collective wellness. To capitalize on this opportunity, cities must strive towards a climax urban state, through the fusion of green infrastructure, localism, and public art.

However, we must ask: to what extent can urban densification be delivered? Will the effort produce a reasonable return on investment? And what role should government have in the enterprise?

Delivering on Densification

The state of California has greatly contributed to the reinvention of the American city: It built the first high-speed highway, the Arroyo Seco Freeway in Pasadena; it showcased the idyll of suburbia by spreading development through hillsides and valleys; and, in response to unbridled sprawl, it launched new urbanism as a model of development. Today, California may well be leading the preference for more compact urban living. Of Gen Y Californians with the intent to move, two-thirds view as important living in a walkable community close to transit, shopping services, and entertainment, while 70 percent do not believe they have to move to the suburbs once they have children. California Gen Y members lead the nation by a significant margin in giving public transit due consideration when choosing a place to live. Fifty percent would support living in a "smart community" as defined by a mix of housing with tree-lined, walkable streets in close proximity to a town center supporting shops, services, and transit; and an equal percentage would live in multistory, multifamily housing in high-density neighborhoods[3] (see figure 10.1).

These statistics bear a direct impact on the housing market. In the sprawling San Diego region, multifamily apartments and townhomes account for nearly 60 percent of the anticipated housing demand through 2035. Statewide, the number rises to 72 percent. In California the demand for density is clearly there—and as California goes, so will much of the rest of the nation. However, the statewide supply of land suitable for higher density development within close proximity of existing and/or planned light rail transit stations can only deliver two-thirds of the demand for denser housing[4]. This calculation is based on an assumed net floor area ratio (FAR) of 2.5. To meet the demand, therefore, an increase in develop-

Figure 10.1. The planned Brisbane Bayland in south San Francisco catpures California's trend towards more compact, transit-oriented urban living. Under development by the Universal Paragon Corporation, the 548-acre project will have upwards of 4,500 dwelling units, more than three-quarters of them flats.

Figure 10.2. A public plaza and rain garden facing the Schuylkill River anchor the southern corner of the Ridge development. Solar panels over green roofs are part of the green systems delivering "net-zero" carbon emissions.

ment intensity will be necessary, possibly to a FAR of 4 or greater. The delivery of such intensity forcibly creates the need for integrated amenities—greenery, places for recreation, public art, etc.

How individual houses are designed also bears relevance to the delivery of density. Cohabitation is a rising trend. More young adults are opting for shared living arrangements, parents and offspring are staying together longer, and extended immigrant families are living in close quarters. Mindful of this trend, the Urban Land Institute is urging residential developers to consider "fewer hallways, smaller bedrooms, larger kitchens opening onto family space, and big living room space for groups to congregate around TV and movie entertainment [along with] club rooms and shared recreation facilities like rooftop pools and workout rooms."[5] Common living areas are gaining more importance as a necessary amenity—both inside living units, as the ULI invites developers to consider, but also as shared outdoor space providing immediate access to air and light, wildlife, views, and street activity.

The Ridge, a "net-zero" (carbon neutral) mixed development proposal in Philadelphia has tapped this trend (see figure 10.2). The project consists of 146 apartments in three- and four-story wings placed above a one-level parking podium lined with ground level retail. A public plaza anchors a busy street corner facing the Schuylkill River. Every unit is single-loaded from exterior corridors to induce energy-saving cross-ventilation. Access is from the lattice-shaded exterior corridors with appended alcoves for resident congregation. Two landscaped interior courtyards provide common areas for outdoor relaxation and social exchange (see figure 10.3).

The developer and architect, Onion Flats, regards the project not as the building of apartments, but rather as the "building of community." The company gained the right to develop the site, which is owned by the Philadelphia Redevelopment Authority (PRA), through a competitive bidding process. As a PRA-sponsored effort, the project must meet the city's mandate for the integration of public art. At a meeting of the PRA's public art committee, various approaches to art were discussed, from real-time interactive displays of green infrastructure measures at work (such as kilowatts generation from rooftop solar panels), to digital murals and streetside seating that would invite the larger neighborhood community to sip coffee while enjoying the river views.

Figure 10.3. Access corridors with lounging alcoves overlook a community courtyard. Dwellings have been omitted at the courtard level to afford views of the river—a gesture that seamlessly fuses inside and outside space as community and environmental assets.

The Ridge project meets the building-as-garden model as a marketable answer to mixed use densification. It also meets the community-as-park model by providing street-side amenities for public gathering and social exchange. As more people are placed in the same acre of urban land, shared public places—sidewalks, plazas, trails, streets, greenways, and parks—will become more highly valued as development assets. Without a linkage to such amenities, individual developments, however green and dense they may be, are destined to become mere islands of sustainability amid a stunted urban ecology. It is the full immersion into a climax urban state that can raise densification beyond the level of individual acceptability to that of collective desirability.

Return on Investment

Initiatives such as the Atlanta Beltline and the Trinity River Corridor Project (TRCP) represent major works of green infrastructure aiming to effect the desirability of core urban areas as denser living and working environments. The 23-mile Atlanta Beltine will tie together 46 neighborhoods through parks, transit, trails, and bikeways, transforming an abandoned rail corridor into a green public loop around the city's core area. As described in Chapter 8, the TRCP is envisioned as the catalyst for new transit- and park-oriented development encompassing close to a square mile of urban land flanking a proposed 2,000-acre urban park. The value of new development spurred by such initiatives is substantial. In the case of Dallas, the value of potential development associated with the TRCP is estimated to be more than six times greater than the cost of the park itself. (The author developed the calculation based on the following: one square mile equals 640 acres; subtracting 50 percent for roadways, easements, and public facilities yields 320 acres available for development; a floor area ratio (FAR) of three yields approximately 42 million square feet of mixed use construction; at $125 per square foot on average, the value of this construction is $5.25 billion. The park improvements are estimated to be approximately $850 million. It should be noted that a FAR of three corresponds to the low end of current development activity within a few blocks of the Trinity River floodway.)

But a green infrastructure return on investment does not derive solely from its development-spurring value. Parks, greenways, and green streets provide substantial return on investment in and of themselves. A study by the National Trust for Public Land suggests that an acre of urban parkland generates approximately $900 in annual economic benefits, accruing from direct usage (fees), improved personal health, tourist visitation, and increased property taxes, among other factors (2009 dollars). (Calculation is based on the average total value per acre derived from the cities listed in Harnik and Welle 2009, using park acreage per city as identified in the Trust for Public Land, Acres of Parkland by City and Agency[6].

In Washington, D.C., the Harnick and Welle's study shows a $1.2 billion bump in real estate value for apartments, condominiums, row houses, and detached homes within 500 feet of a park, yielding nearly $7 million in taxes. In San Diego, Mission Bay Park and Balboa Parks, the city's two major recreation and cultural destinations, respectively, yield nearly $40 million annually in tourist expenditures. And in Boston, more than $354 million is generated annually through the use of the city's 5,000 acres of parkland for active recreation, organized sports, festivals, concerts, and attractions.

Parks, of course, are repositories of canopy trees that sequester carbon, stabilize slopes, mitigate the heat island effect, and moderate stormwater flows. The Trinity River Corridor will have 19,000 bottomland trees whose carbon absorption potential is estimated to be 380 tons per year, the equivalent of 750,000 vehicular miles traveled.

Trees anywhere in urban areas can perform such infrastructural service. A study of street trees in New York City found that the climate-moderating benefits provided by trees result in annual energy savings of $27.8 million, or $47.63 per tree.[7] The National Tree Benefit Calculator identifies the monetary benefit of trees according to zip code, species, size, and land use. For example, a 24-inch caliper red maple in Philadelphia represents a $200 annual value within a multifamily residential area. This value is obtained from the detention, through foliage and roots, of 8,000 gallons of stormwater; from the conservation of 200 kilowatts per hours by shading building walls; from the absorption of 1,000 pounds of CO_2 through the tree's biomass; and from an increase in property values, among other factors[8].

Along with rain gardens, porous paving, bioswales, and cisterns, street trees will help Philadelphia meet the EPA-approved conversion of one-third of the impervious cover within the city's combined-sewer watershed into "greened-acres" (9,564 acres, to be exact).[9] The greened-acres program is part of the city's agreement with the EPA to meet the mandates of the Clean Water Act. It was derived in part by a cost-benefit analysis of a "green" versus "gray" approach to stormwater management. "Green" yielded greater overall benefits, including: six percent greater visitation to Fairmount Park; a 3.5 percent increase in property values; creation of 380 green infrastructure related jobs; 160 fewer fatalities caused by excessive heat events; and 250 fewer days of school absence due to respiratory ailments.[10]

It is safe to conclude that urban environments developed under a "greened-acres" program, such as Philadelphia's, would be more desirable places to live and work than areas lacking the same degree of amenity. People in such environments would lead healthier lives, walking more, exercising more, and breathing cleaner air. The need for health care would be lessened, as would energy consumption and the need for expensive, "gray" stormwater control systems. Such cost savings can be construed as a form of "green equity," an economic surplus that can also be made available for other types of investments, such as social services and public education.

A Role for Government

Cobbling together funds for initiatives such as the Atlanta Beltline and the Trinity River Corridor Project requires sustained political will, devoted civic leadership, willing donors, and the accommodation of a voting public—conditions highly susceptible to change, especially on projects requiring phases of implementation over multiple economic and election cycles. In Atlanta, close to 60 percent of the $3 billion Beltine cost is anticipated to be financed through tax allocation districts (a form of tax increment financing) covering more than ten square miles of adjoining land. Another $300 million is further anticipated from federal and state sources, nonprofit partners and donors. The city sought supplementary financing through a 2012 referendum on state-wide transportation improvements financed through a one-cent sales tax increase. On July 31, 2012, voters rejected the appeal for this financing. Whether or not full financing for the initiative will be obtained remains uncertain.

In the case of Dallas, one-third of the funds for the $2 billion Trinity River project (including the Trinity Parkway) have been obtained. Funding sources include a voter-approved improvement bond, the federal government's Water Allocation Resources Act, the North Texas Transportation Authority, the U.S. Army Corps of Engineers, plus several millions from private donors. However, after more than 15 years of planning and design, full funding for the project remains in doubt.

One success story is the Parklands of Floyds Fork in Louisville, Kentucky (see Chapter 9).[11] The $120 million, 18-mile greenway is well under construction after garnering federal, state, city, and donor funds over the course of a decade. The initiative materialized through the philanthropic vision and leadership of the Jones family (David, head of the Humana Corporation, and his son Dan, head of 21st Century Parks, the nonprofit organization set up to manage the effort). Approximately three-quarters of the project costs, including land acquisition, have been privately funded. It is extraordinary that the nation's largest urban park is being realized through philanthropic means and funded to such a great extent, to include future maintenance, by private donations.

However, philanthropy cannot be counted on as a means to leverage development and move the nation towards a climax urban state. To "weatherproof" the nation, a more predictable, dependable, and effective method is necessary, such as occurred with the Interstate Highway System more than half a century ago. A connected, military-capable network of roadways linking every major city, crisscrossing the country north to south, east to west; and overcoming mountains, deserts, and a noted "river of grass" would not have occurred by tapping the goodwill of disparate donors, narrow corporate interests, or localized political agendas. In other words, the transformation of the nation's cities into low energy, diverse, resilient, and healthy "climax" environments should be viewed as a matter of national security every much as vital as creating a system of highways by which to mobilize armament and place the population out of harm's way.

When Dwight D. Eisenhower assumed the presidency, there was little time to waste in implementing his long-standing vision of a nation connected coast-to-coast by highways. His secretary of the interior was Douglas McKay, owner of a Portland auto dealership; his treasury secretary was George M. Humphry, former president of the M.A. Hanna Steel company and future president of the National Steel Corporation; his commerce secretary was Sinclair Weeks, a metal products manufacturer; and Martin P. Durkin, his labor secretary, was past president of the United Association of Plumbers and Steamfitters. Under such a brain trust, the economic benefits of highways easily melded with a national defense agenda. Levittown on Long Island had been completed a few years earlier, a blueprint towards homeownership for millions of Americans. New highways would facilitate the spread of such developments throughout the land. Automobiles, made of steel and assembled under secured pacts with labor unions, would provide front door access. In combination, steel, roads, cheap land, home building, and automobiles produced a lasting economic boom. The Interstate highways provided key leverage.

Begun in 1956, the 42,800-mile system is regarded as the greatest public work in the nation's history, if not the world. Its ultimate cost is estimated at $510 billion in 2012 dollars.[12] Ninety percent of this cost has been borne by the federal government, paid through the Highway Trust Fund (HTF). Established in 1956, the HTF is replenished through taxation, primarily of gasoline and motor oils. The US Congress periodically re-authorizes the use of the HTF through surface transportation acts.

The latest authorization, the Moving Ahead for Progress in the Twenty-First Century Act (MAP-21), signed into law by President Obama on July 6, 2012, provides $101 billion in funding through Fiscal Year 2014. The act's Transportation Alternatives Program provides money for greenways, trails, and other active mobility initiatives, including safe routes to schools, but to a very limited extent—only two percent of the bill's funding allocation. Another 9.6 percent is directed toward public transportation.[13]

These percentages are a mismatch with regards to national productivity. The efficiency gained in the transport of goods from the interstate highway system as a whole caused a spike in national productivity—more than 30 percent during the 1950s. But, as the system matured, the gains declined, to seven percent as early as the 1980s.[14] No new federal highways are currently contemplated. The system's function as a land development catalyst has substantially diminished. Much of today's highway bill funding is directed toward repair and maintenance operations and localized roadway widening designed to alleviate congestion. Redirecting the allocation of gasoline and motor oil taxation to other avenues for growth and development would be a rational course, especially if it delivered a reasonable return on investment. A climax cities program would deliver such value, generating an urban densification-based economy.

If national defense is an incontestable federal responsibility, then surely the development of energy-efficient cities that facilitate active mobility and public transit, that encourage healthier, active lifestyles, that conserve water and improve air quality, and that are resilient in the face of a changing climate deserves greater and more consistent federal support. If the dispersal of the population was once deemed a defense strategy against nuclear attack, then its concentration in quality urban environments must be regarded as today's defense imperative. Augmenting the mission of the HTF to embrace the full gamut of urban green infrastructure measures and increasing the funding allocation towards them would achieve this end.

The choice is clear: stay the course and slowly gain on green densification, or increase the rate by which nonrenewable energy is conserved, water harvested and recycled, CO_2 sequestered, obesity reduced, wars avoided. The sustainable densification of the nation's cities can proceed on a catch-as-catch-can basis subject to economic and public voting cycle uncertainties, or it could do so as a national priority, with an implementation approach based on the principle of maximum long-term accrued green equity.

On November 2, 2011, President Obama stood on a podium erected at the labyrinth in the Georgetown Waterfront Park and delivered a speech in support of the American Jobs Act, making specific mention of the need to fund the repair of the nation's aging roadway infrastructure. The Key Bridge over the Potomac River stood strategically in the background. A few yards from the gathered crowd, one of the park's largest rain gardens and its longest bioengineered river revetment went unnoticed, obscured by an oversized American flag.

The president's plea was not unfounded, of course. As the oldest surviving bridge in Washington, D.C., the Key Bridge is indeed in need of a major overhaul. Initially conceived as a connector between Georgetown and Rosslyn across the Potomac River, the bridge today provides access to communities much farther afield via Interstate Highway 66. Past Rosslyn and downtown Arlington the transition to the suburban landscape of northern Virginia is almost immediate. There the subway emerges from its tunnel and enters the median of the interstate for nine miles to the Vienna station, last stop within sprawling Fairfax County. Thereafter, the sight of multistory buildings is sporadic, an unexpected sight within a development grain that generally stays below the treetops. Nine miles further, past Centreville and the last exit to Dulles International Airport, farmland emerges to complete the exurban mosaic. Another 12 miles out, past the village of Haymarket, the last single-family subdivision is left behind.

A 30-mile arc, more or less, from the nation's capital defines the limits of suburbia within its sphere of influence. More than 5.6 million people live within this arc, one of the nation's fastest growing metropolitan areas. Many of them face daily hour-long automobile commutes to work. There are plans to extend the Washington Metro from the Vienna station to Centreville to ease automobile dependency, but they are a low priority and remain unfunded. In Montgomery County, Maryland, the last suburban outpost within the Interstate 270 growth corridor is Clarksburg, 27 miles from the center of the nation's capital. Bus rapid transit is planned from this small town to Shady Grove, the nearest Washington metro station. The 15-mile run is estimated to cost $800 million.[15] It will deliver a 50-minute bus ride to the station; thereon, commuters will spend another 40 on the metro before reaching downtown Washington.

One has to wonder: At what point does a 90-minute transit-assisted commute stop making sense? What, then, to do with the national suburban morass to which extending public transit will never be cost effective? What, then, to do with the nation's insurmountable highway and bridge upgrade and repair bill, a constant erosion to the Gross National Product that perpetuates an unsustainable pattern of living?

Perhaps a future presidential cabinet will include a different mix of occupational backgrounds—individuals who have risked fame and fortune developing green multifamily urban housing who have expertise in ecological economics and eco-system services; who have pioneered urban farming; who have led research on obesity and public health, who have instituted green infrastructure programs as a way to improve the quality of the nation's waters, and who have made a career of promoting public art as part of the creative economy. Perhaps then a future administration will choose to explain to mil-

BEYOND, AHEAD

lions of Americans that the world has changed, that a highway-driven prosperity is a thing of the past and that, along with it, the dream of a secure home in a healthy environment no longer depends on a bucolic suburban setting but, rather, amid the vibrancy and artful greenery of a climax city.

A FINAL WORD ON IAN L. MCHARG

The man spoke powerfully, effectively and incessantly on the need to save the environment from abuse and despoliation in the hands of uninformed developers and enabling public bureaucracies. In doing so he may have contributed to the spread of development well into the hinterland. Through this book I have challenged this and other aspects of his legacy, but the call to build a healthier and safer world protective of our most precious environmental resources remains as valid today as when *Design with Nature* was first published. Today it is cities, rather than the hinterland, that are the precious resource.

Cities constitute our natural milieu and, as such, they must be qualitatively elevated from a landscape and building standpoint. Planners, urban designers, architects, landscape architects, artists, engineers, and a host of related experts must come together to effect such elevation. A Unity of the Arts must be pursued and achieved. And, as McHarg preached, ecology must remain the common theme, systematically applied. But this now must be a complete ecology—a Leibnizian ecology—that, through art, fuses matter (green) and spirit (localism) into building blocks for our earthly abode.

Perhaps this is what Alan Berger had in mind when he wrote:

> How does McHarg's model for ecological planning evolve? Even the great projects based on McHargian theory—the Woodlands of the world—have been encroached by short-term, quick-gain market forces and mentalities. Facing off against, or perhaps slowing down these forces, is not what Ian had on mind when he spoke of putting 'together again the entire system.' He was, instead, brilliantly laying out the next great project for landscape architects: to pick up the pieces of regional systems left in the wake of economic schema, political indecision, ad hoc development, a negligent public and flawed environmental health policy.[16]

End Notes

Foreword

1. Julian P. Boyd et al., ed. *The Papers of Thomas Jefferson* (Princeton: PrincetonUniversity Press, 1950). Volume 1, Chapter 18, Document 21.
2. Thomas Jefferson, "Notes on the State of Virginia" (1781to 1785). Reprinted in Andrew A. Lipscomb, ed. *The Writings of Thomas Jefferson*, vol. 2, 230.
3. www.planning.org/conference/previous/2011/coverage/openingkeynote.htm?print=true.

Chapter 1

1. Ian L. McHarg, *Design with Nature* (New York: Natural History Press, 1969), page 5.
2. Wallace-McHarg Associates, *Plan for the Valleys*. Report for the Valleys Planning Association, 1963.
3. McHarg, 1969, 120.
4. Ibid., 3.
5. Wallace-McHarg Associates, 1963, Preamble.
6. David A. Wallace, *Urban Planning/My Way: From Baltimore's Inner Harbor to Lower Manhattan and Beyond* (Washington, D.C.: Planners Press, 2004), 78.
7. McHarg, 1969, 34.
8. Raymond A. Mohl, *The Interstates and the Cities: Highways, Housing, and the Freeway Revolt* (Washington, D.C.: Poverty and Race Research Action Council, 2002), 30.
9. Ian L. McHarg, *A Quest for Life* (New York: John Wiley & Sons, Inc., 1996), 2.
10. Gerard K. O'Neil, *The High Frontier: Human Colonies in Space* (New York: William Morrow and Company, 1977).
11. Bill McKibben, *The End of Nature* (New York: Anchor, 1989).

12. Katheryn Hayles, "Simulated Nature and Natural Simulations: Rethinking the Relationship between the Beholder and the World," in *Uncommon Ground: Rethinking the Human Place in Nature*, ed. William Cronon (New York: W.W. Norton & Co., 1995)

Chapter 2

1. William H. MacLeish, *The Day before America: Changing the Nature of a Continent* (New York: Houghton Mifflin Co., 1994).
2. MacLeish, 1994, 115.
3. Alvar Nuñez Cabeza de Vaca, La Relación, Trans: Martin A. Favata and José B. Fernández (Houston: Arte Público Press, 1993), 54.
4. Roderick Nash, *Wilderness and the American Mind*, 4th ed. (New Haven and London: Yale University Press, 2001).
5. Ian L. McHarg, *A Quest for Life* (New York: John Wiley & Sons, Inc., 1996), 183-189.
6. U.S. Geological Survey. *Geologic Provinces of the United States: Basin and Range Province*. http://geomaps.wr.usgs.gov/parks/province/basinrange.html
7. Carson, U.S. Fish and Wildlife Service website.

Chapter 3

1. Ruben N. Lubowski, et al., *Major Uses of Land in the United States*, 2002, 2005. U.S. Department of Agriculture, Economic Information Bulletin, No. 14, 28.
2. Lawrence D. Franks, Martin A. Andresen and Thomas L. Schmid, "Obesity Relationships with Community Design, Physical Activity, and Time Spent in Cars." *American Journal of Preventive Medicine*, August, 2004. Vol. 27. Issue 2, 87-96.
3. Howard Frumkin, Lawrence Franks and Richard Jackson. *Urban Sprawl and Public Health: Designing, Planning and Building for Healthy Communities* (Washington, D.C.: Island Press, 2004).
4. Peter Calthorpe, et al., *The Ahwahnee Principles for Resource Efficient Communities*. (Sacramento, California: Local Government's Commission Center for Livable Communities, 1991).
5. www.lgc.org/ahwahnee/principles.html
6. Wallace Roberts & Todd, *South Livermore Specific Plan*, 1997. Report to the City of Livermore, California.
7. Rob Aldrich and Christina Soto, The 2010 National Land Trust Census Report, 2011. Land Trust Alliance, 5.
8. The Valleys Planning Council, Easements Program, www.thevpc.org/the-valleys-today/easement-programs/
9. 2000 U.S. Census; www.infoplease.com/us/census/data/

End Notes

Chapter 4

1. Ian L. McHarg, *Design with Nature* (New York: Natural History Press, 1969) 196.
2. Ian L. McHarg and Frederick Steiner, *To Heal the Earth* (Washington, D.C.: Island Press, 1998), 35.
3. Anne Whiston Spirn, *The Granite Garden: Urban Nature and Human Design* (New York: Basic Books, 1984).
4. Ibid., xi.
5. McHarg and Steiner, 1998, 215.
6. Ibid., 217.
7. Anne Whiston Spirn, "Restoring Mill Creek: Landscape Literacy, Environmental Justice and City Planning and Design," *Landscape Research* 30, No. 3, 2005, 395-413.
8. Interview with Anne Whiston Spirn, University of Pennsylvania Gazette, Mar/Apr 2002.
9. www.phila.gov/green/greenworks
10. Wallace Roberts & Todd, *GreenPlan Philadelphia: Our Guide to Achieving Vibrant and Sustainable Urban Place*. Report to the City of Philadelphia, 2010, 9.
11. Ibid., 34.
12. EPA Office of Transportation and Air Quality, *Emission Facts: Greenhouse Gas Emissions from a Typical Passenger Vehicle*, 2005.

Chapter 5

1. Robert Harbison, *Reflections on Baroque* (Chicago: University of Chicago Press, 2000), 8.
2. Gilles DeLeuze, *The Fold, Leibniz and the Baroque*, Trans. Tom Conley (Minneapolis: University of Minnesota Press, 1993), 123.
3. Ibid., 1993, xvi.
4. Ernst Cassirer, *The Philosophy of the Enlightenment* (Princeton: Princeton University Press, 1951), 29-30.
5. Charles Waldheim, *Landscape Architecture Reader* (New York: Princeton Architectural Press, 2006).
6. Ignacio F. Bunster-Ossa, "Landscape Urbanism," *Urban Land* 60, no. 7, 38.
7. Dean Almy, ed. 2007, "On Landscape Urbanism," *Center: A Journal for Architecture in America*, No. 14
8. See, for example: Mostafavi and Doherty, ed.. *Ecological Urbanism*, 2010; Beatly, *Green Urbanism*, 2000; Farr, *Sustainable Urbanism*, 2008. (See Bibliography for full citations.)
9. Mostafavi and Doherty, 2010, 528.

Chapter 6

1. Colin Woodard, *American Nations: A History of the Eleven Rival Regional Cultures of North America* (New York: Viking Press, 2011), 7-12.

2. Lucy R. Lippard, *The Lure of the Local: Senses of Place in a Multicentered Society* (New York: New Press, 1997), 196.

3. Xavier De Souza Briggs, *Democracy and Problem Solving: Civic Capacity in Communities Across the Globe* (Cambridge, MA: The MIT Press, 2008) 42.

4. Ibid., 43.

5. Lippard, 2008, 33.

Chapter 7

1. Americans for the Arts, undated, 1.

2. Review of mission statements of the University of California Berkeley, Harvard University, University of Pennsylvania, and University of Virginia.

3. Suzanne Lacy, ed., *Mapping the Terrain: New Genre Public Art* (Seattle WA: Bay Press, 1995), 140.

4. Ian L. McHarg, *Design with Nature* (New York: Natural History Press, 1969), 165.

5. Lacy, 1995, 140.

6. Robert Irwin, *Being and Circumstance: Notes Toward a Conditional Art* (Larkspur Landing, Calif.: The Lapis Press, 1985), 27.

7. Ibid., 77.

8. Sue Spaid, *Ecovention*, Exhibition at The Contemporary Art Center, Cincinnati, 2002.

9. *Interview: Mel Chin on Revival Field*, University of Michigan. www.msu.edu/course/ha/491/melchin.pdf, 397.

10. Mel Chin on Revival Field (1990), from the website *Global Positioning: Exploring Contemporary World Art*, www.melchin.org/oeuvre/artist-writing-revival-field.

11. *The Poetic Utility* plan for Seattle, www.bustersimpson.net/poeticutility/poeticutility-bustersimpson1998.pdf

12. Lorna Jordan Studio. *A Theater of Regeneration: Towards an Aesthetic of Layers, Loops and Lenses?* Broward County Environmental Art Master Plan, 2004, 25.

13. Ibid., 14.

14. Todd Bressi, "EDRA/Places Awards 2004." *Places*, October 2004. http://places.designobserver.com/media/pdf/Theater_of_Reg_661.pdf

Chapter 8

1. James Pratt, *Dallas Visions for Community: Toward a 21st Century Urban Design* (Dallas: Dallas Institute of Humanities and Culture, 1992), Foreward.

2. Antonio Machado, "Last Night As I Was Sleeping," partial lines, second stanza.

3. James Pratt, 1992, 15, 16.

End Notes

4. Ivan Illich, *H₂O and the Waters of Forgetfulness* (Berkeley: Heyday Books, 1985), 7.

5. Ibid., 7.

6. John Williams Rogers, *Lusty Texans of Dallas* (New York: E.P. Dutton and Company, Inc., 1951), 25.

7. Vision North Texas, *Vision North Texas: Regional Choices for North Texas*, 2008.

Chapter 9

1. See, for example Daniel B. Botkin, *Discordant Harmonies* (Oxford: University Press, 1990).

2. Ian L. McHarg, *Design with Nature* (New York: Natural History Press, 1969), 181.

3. Donald Appleyard and Kevin Lynch, *Temporary Paradise?: A Look at the Special Landscape of the San Diego Region*. A report to the City of San Diego, 1974. Available from the Cambridge Department of Urban Studies and Planning, Massachusetts Institute of Technology.

4. The Cultural Landscape Foundation, *City Shaping: The Olmsteds & Louisville*. http://tclf.org /Louisville

5. See the Parklands of Floyds Fork, www.theparklands.org

6. FEMA: http://www.fema.gov/pdf/hazard/flood /recoverydata/orleans_parish04-12-06.pdf

7. Central Park Conservancy, *History of Central Park*.

8. Arthur C. Nelson, *The New California Dream: How Demographic and Economic Trends May Shape the Housing Market; A Land Use Scenario for 2020 and 2035* (Washington D.C.: Urban Land Institute, 2011), 26.

9. North Central Texas Council of Governments,www.nctcog.org/trans/sustdev/bikeped/veloweb.asp

10. Indianapolis Cultural Trail, www.indyculturaltrail.org/about.html; www.indyculturaltrail.org/media _release42811.html

11. William Curtis, *Modern Architecture Since 1900*. 2nd ed. (Englewood Cliffs N.J.: Prentice-Hall, Inc., 1987), 402.

12. www.yangsquare.com/menara-mesiniaga-in-detail/

13. James Wines, *Green Architecture* (Los Angeles: Taschen, 2000), 8.

14. Friedensreich Hundertwasser, *Mouldiness Manifesto against Rationalism in Architecture*, 1964.

Chapter 10

1. Thomas L. Friedman, "One for the Country," *New York Times*, April 17, 2012.

2. Ethan Seltzer, and Armando Carbonell, editors, *Regional Planning in America: Practice and Prospect* (Cambridge, Mass.: Lincoln Institute of Planning Policy), 251.

3. Arthur C. Nelson, *The New California Dream: How Demographic and Economic Trends May Shape the Housing Market; A Land Use Scenario for 2020 and 2035* (Washington D.C.: Urban Land Institute, 2012), 29-36.

4. Ibid., 11.

5. Jonathan Miller, *What's Next: Real Estate in the New Economy* (Washington, D.C.: Urban Land Institute, 2011), 35.

6. Peter Harnick and Ben Welle, *Measuring the Economic Value of a City Park System*. Report by the Trust for Public Land, 2009. www.cloud.tpl.org/pubs/ccpe-econvalueparks-rpt.pdf.

7. American Society of Landscape Architects, the Lady Bird Johnson Wildflower Center at The University of Texas at Austin, and the United States Botanic Garden. *The Sustainable Sites Initiative: Guidelines and Performance Benchmarks*, 2009, 67.

8. National Tree Benefit Calculator. www/treebenefits.com/calculator/

9. City of Philadelphia, *Green City, Clean Waters: Implementation and Adaptive Management Plan; Consent Order & Agreement, Deliverable I*, Report Submitted to the Commonwealth of Pennsylvania Department of Environmental Protection, December 1, 2011.

10. City of Philadelphia, *Combined Sewer Overflow Long Term Control Plan Update*, October 1, 2009.

11. Wallace Roberts & Todd. *The Fork: The Floyd's Fork Greenway Master Plan*. Report to 21st Century Parks, 2008.

12. http://en.wikipedia.org/wiki/Interstate_Highway_System; calculation of inflation: inflationdata.com /Inflation/Inflation_Rate/HistoricalInflation.aspx)

13. National Cooperative Highway Research Program (NCHRP), Project 20-24 (52), Future Options for the National System of Interstate and Defense Highways: Technical Memorandum Task 2, The Economic Impact of the Interstate Highway System. 2006. www.interstate50th.org/docs/techmemo2.pdf

14. National Cooperative Highway Research Program (NCHRP), Project 20-24 (52), Technical Memorandum Task 2.

15. Maryland Department of Transportation Announcement. www.cctmaryland.com/images/stories /documents.lpa.CCT Alternative Announcement Press Release.pdf

17. Lynn Margulis, James Corner and Brain Hawthorne, ed., *Ian McHarg Conversations with Students: Dwelling in Nature* (New York: Princeton Architectural Press, 2007), 8.

Bibliography

Aldrich, Rob, and James Wyerman. 2006. *The 2005 National Land Trust Census Report.* Land Trust Alliance. www.northolympiclandtrust.org/Documents/2005LandTrustCensusReport.pdf

Almy, Dean, ed. 2007. "On Landscape Urbanism." *Center: A Journal for Architecture in America* 14.

American Society of Landscape Architects, the Lady Bird Johnson Wildflower Center at the University of Texas at Austin, and the United States Botanic Garden. 2009.

Appleyard, Donald, and Kevin Lynch. 1974. "Temporary Paradise: A Look at the Special Landscape of the San Diego Region." A report to the City of San Diego.

Beatley, Timothy. 2004. *Green Urbanism: Learning from European Cities.* Washington, D.C.: Island Press.

Biourbanism, International Society of: *What is biourbanism.* www.biourbanism.org/

Botkin, Daniel B. 1990. *Discordant Harmonies.* Oxford University Press.

Bressi, Todd. 2004. EDRA/Places Awards 2004. *Places.* October: 22. Available at http://places.designobserver.com/media/pdf/Theater_of_Reg_661.pdf

Bunster-Ossa, Ignacio F. 2001. "Landscape Urbanism." *Urban Land* 60 (7).

Cabeza de Vaca, Alvar Núñez. 1993. *La Relación.* Trans. Martin A. Favata and José B. Fernández. Houston: Arte Público Press. Available at http://alkek.library.txstate.edu/swwc/cdv/la_relacion/index.html.

Carson, Rachel, U.S. Fish and Wildlife Service website, www.fws.gov/refuge/Rachel_Carson/about/rachelcarsonexcerpts.html. The essay "History of the National Wildlife Refuge System" introduced the series, "Conservation in Action," a collection of narratives about refuges and the refuge system. When she wrote this, Carson was a scientist and the chief editor for the U.S. Fish and Wildlife Service.

Calthorpe, Peter, et al. 1991. *The Ahwahnee Principles for Resource Efficient Communities.* Local Government Commission. www.lgc.org/ahwahnee/ahwahnee_principles.pdf

Cassirer, Ernst. 1951. *The Philosophy of the Enlightenment.* Princeton, N.J.: Princeton University Press.

Central Park Conservancy. *History of Central Park.* Available at www.centralparknyc.org/visit/history/

RECONSIDERING IAN MCHARG

Cronon, William, ed. 1995. *Uncommon Ground: Rethinking the Human Place in Nature*. New York: W.W. Norton & Co.

Cultural Landscape Foundation. *City Shaping: The Olmsteds & Louisville*. http://tclf.org/louisville

Curtis, William. 1987. *Modern Architecture Since 1900*. 2nd ed. Englewood Cliffs, N.J.: Prentice-Hall, Inc.

DeLeuze, Gilles. 1993. *The Fold, Leibniz and the Baroque*. Trans. Tom Conley. Minneapolis: University of Minnesota Press.

De Souza Briggs, Xavier. 2008. *Democracy and Problem Solving: Civic Capacity in Communities Across the Globe*. Cambridge, Mass.: The MIT Press.

Emo Urbanism. http://en.wikipedia.org/wiki/Charles_Morris_Anderson. Also, KF Foundation. http://naturesacred.org/

Farr, Douglas. *Sustainable Urbanism: Urban Design with Nature*. 2008. Hoboken, N.J.: John Wiley and Sons.

Franks, Lawrence D., Martin A. Andersen, and Thomas L. Schmid. 2004. "Obesity Relationships with Community Design, Physical Activity, and Time Spent in Cars." *American Journal of Preventive Medicine* 27 (2): 87-96. Available at www.ajpmonline.org/article/S0749-3797(04)00087-X/fulltext

Friedman, Thomas L. 2012. "One for the Country." *New York Times*, April 17, Opinion Pages. Available at www.nytimes.com/2012/04/18/opinion/friedman-one-for-the-country.html?ref=thomaslfriedman

Frumkin, Howard, Lawrence Franks and Richard Jackson. 2004. *Urban Sprawl and Public Health: Designing, Planning and Building for Healthy Communities*. Washington, D.C.: Island Press.

Harbison, Robert. 2000. *Reflections on Baroque*. Chicago: University of Chicago Press.

Harnik, Peter and Ben Welle. 2009. *Measuring the Economic Value of a City Park System*. Report by the Trust for Public Land. Available at http://cloud.tpl.org/pubs/ccpe-econvalueparks-rpt.pdf

Historical Inflation Rate. http//inflationdata.com/Inflation/Inflation_Rate/HistoricalInflastion.aspx

Hundertwasser, Friedensreich. 1958/1959/1964. *Mouldiness Manifesto Against Rationalism in Architecture*. Available at www.hundertwasser.at/english/texts/philo_verschimmelungsmanifest.php

Illich, Ivan. 1985. H_2O *and the Waters of Forgetfulness*. Berkeley: Heyday Books.

Indianapolis Cultural Trail. www.indyculturaltrail.org/

Interview: Mel Chin on Revival Field, University of Michigan. https://www.msu.edu/course/ha/491 /melchin.pdf, p. 397

Irwin, Robert. 1985. *Being and Circumstance: Notes Toward a Conditional Art*. Larkspur Landing, Calif.: The Lapis Press.

Kwon, Miwon. 2002. *One Place after Another: Site-Specific Art and Locational Identity*. Cambridge, Mass.: The MIT Press.

Lacy, Suzanne, ed. 1995. *Mapping the Terrain: New Genre Public Art*. Seattle: Bay Press.

Lippard, Lucy R. 1997. *The Lure of the Local: Senses of Place in a Multicentered Society*. New York: New Press.

Bibliography

Lorna Jordan Studio. 2004. *A Theater of Regeneration: Towards an Aesthetic of Layers, Loops and Lenses; Broward County Environmental Art Master Plan*. Available at www.broward.org/Arts/Resources/Publications/Documents/master_plan_parks_bond.pdf

Lubowski, Ruben N., et al., 2005. *Major Uses of Land in the United States, 2002*. U.S. Department of Agriculture Economic Information Bulletin Number 14, 2005. www.ers.usda.gov/media/249900/eib14fm_1_.pdf

MacLeish, William H. 1994. *The Day before America: Changing the Nature of a Continent*. New York: Houghton Mifflin Co.

Margulis, Lynn, James Corner, and Brain Hawthorne, ed. 2007. *Ian McHarg Conversations with Students: Dwelling in Nature*. New York: Princeton Architectural Press.

McHarg, Ian L. 1969. *Design with Nature*. New York: Natural History Press.

McHarg, Ian L. 1996. *A Quest for Life*. New York: John Wiley & Sons, Inc.

McHarg, Ian L. and Frederick Steiner. 1998. *To Heal the Earth*. Washington, D.C.: Island Press.

McKibben, Bill. 1989. *The End of Nature*. New York: Anchor

Maryland Department of Transportation Announcement. www.cctmaryland.com/images/stories/documents/lpa/CCT_Alternative_ Announcement_Press_Release.pdf

Mel Chin on *Revival Field* (1990), from the website *Global Positioning: Exploring Contemporary World Art*, 2003, Chin 2003; also, http://melchin.org/oeuvre/artist-writing-revival-field

Mel Chin Quote. http://greenmuseum.org/c/aen/Issues/chin.php

Mohl, Raymond A. 2002. *The Interstates and the Cities: Highways, Housing, and the Freeway Revolt*. Poverty and Race Research Action Council, www.prrac.org/pdf/mohl.pdf

Miller, Jonathan. 2011. *What's Next: Real Estate in the New Economy*. Washington, D.C.: Urban Land Institute.

Mostafavi, Mohsen and Gareth Doherty, eds. 2010. *Ecological Urbanism*. Baden, Switzerland: Lars Muller Publishers.

Nash, Roderick. 2001. *Wilderness and the American Mind*. 4th ed. New Haven and London: Yale University Press.

National Cooperative Highway Research Program (NCHRP). 2006. Project 20-24 (52), *Options for the National System of Interstate and Defense Highways: Technical Memorandum Task 2, The Economic Impact of the Interstate Highway System*. Available at www.interstate50th.org /docs/techmemo2.pdf.

National Tree Benefit Calculator. www.treebenefits.com/calculator/

Nelson, Arthur C. *The New California Dream: How Demographic and Economic Trends May Shape the Housing Market; A Land Use Scenario for 2020 and 2035*. 2011. Washington, D.C.: Urban Land Institute.

New Orleans, City of. *The Unified New Orleans Plan: Citywide Strategic Recovery and Rebuilding Plan*. 2006. Available at http://quake.abag.ca.gov/wp-content/documents/resilience/New%20Orleans-FINAL-PLAN-April-2007.pdf

North Central Texas Council of Governments. *Regional Veloweb*. www.nctcog.org/trans/sustdev/bikeped/veloweb.asp

O'Neil, Gerard K. 1977. *The High Frontier: Human Colonies in Space*. New York: William Morrow and Company.

Philadelphia, City of. 2009. *Combined Sewer Overflow Long Term Control Plan Update*. October 1. Available at www.phillywatersheds.org/ltcpu/Vol02_TBL.pdf

———. 2011. *Green City, Clean Waters: Implementation and Adaptive Management Plan; Consent Order & Agreement, Deliverable I*. Report Submitted to The Commonwealth of Pennsylvania Department of Environmental Protection. December 1.

Pratt, James. 1992. *Dallas Visions for Community: Toward a 21st Century Urban Design*. Dallas: Dallas Institute of Humanities and Culture.

Public Art Network Council. *Why Public Art Matters*. Green Paper. Undated. http://blog.artsusa.org/artsblog/wp-content/uploads/greenpapers/documents/PublicArtNetwork_GreenPaper.pdf

Rouse, David and Ignacio F. Bunster-Ossa. 2013. *Green Infrastructure: A Landscape Approach*. Planning Advisory Service Report 571. Chicago: American Planning Association,

Safdie, Moshe. 1974. *For Everyone a Garden*. Cambridge, Mass.: The MIT Press.

Seattle Public Utilities, www.seattle.gov/util/)art master plan, Poetic Utility, http://www.bustersimpson.net/poeticutility/poeticutility-bustersimpson1998.pdf.

Seltzer, Ethan, and Armando Carbonell, ed. 2011. *Regional Planning in America: Practice and Prospect*. Cambridge, Mass.: Lincoln Institute of Land Policy.

SITE. *Architetcture as ART*. 1980. New York: St. Martin's Press.

Spaid, Sue. 2002. *Ecovention*. Cincinnati: The Contemporary Art Center.

Spirn, Anne Whiston. 1984. *The Granite Garden: Urban Nature and Human Design*. New York: Basic Books.

———. 2005. "Restoring Mill Creek: Landscape Literacy, Environmental Justice and City Planning and Design." *Landscape Research*. 30 (3) July, 395-413.

Steiner, Frederick, ed. 2006 (first published in 1998). *The Essential Ian McHarg: Writings on Design with Nature*. Washington, D.C.: Island Press.

———. 2008. *The Living Landscape*, 2nd ed. Washington, D.C.: Island Press. 2008.

———. 2011. *Design for a Vulnerable Planet*. Austin, Tex.: The University of Texas Press.

Stoneman Douglas, Marjory. 2007. *The Everglades: River of Grass*. 60th anniversary edition. Sarasota, Fla.: Pineapple Press.

U.S. Department of Agriculture, Economic Research Services. 2006. *Major Uses of Land in the United States*. Economic Information Bulletin No. (EIB-14). May. Available at www.ers.usda.gov/publications/eib-economic-information-bulletin/eib14.aspx.

Bibliography

U.S. Department of Homeland Security, Federal Emergency Management Agency. *Flood Recovery Guidance: Advisory Base Flood Elevations for Orleans Parish, Louisiana*. www.lalandtrust.us/RFP/App_G_ABFE_Guidance_(FEMA_Publication)_for_Orleans_Parish_04-12-06.pdf

U.S. Department of Transportation, Federal Highway Administration. *Frequently Asked Questions: What Did It Cost?* Available at www.fhwa.dot.gov/interstate/faq.htm#question6.

———. *MAP 21*. Available at www.fhwa.dot.gov/map21/.

U.S. Environmental Protection Agency, Office of Transportation and Air Quality. 2005. *Emission Facts: Greenhouse Gas Emissions from a Typical Passenger Vehicle*. EPA 420-05-04.

U.S. Fish & Wildlife Service. *History of the National Wildlife Refuge System*. Available at www.fws.gov/refuges/history/over/over_main_fs.html

Vision North Texas. 2008. *Vision North Texas: Regional Choices for North Texas*.

Waldheim, Charles. 2006. *Landscape Architecture Reader*. New York: Princeton Architectural Press.

Wallace, David A. 2004. *Urban Planning/My Way: From Baltimore's Inner Harbor to Lower Manhattan and Beyond*. Washington, D.C.: American Planning Association Planners Press.

Wallace-McHarg Associates. 1963. *Plan for the Valleys*. Report for the Valleys Planning Association, 1963.

Wallace Roberts & Todd. 1967. *South Livermore Specific Plan*. Report to the City of Livermore, California. November.

———. 2005. *Action Plan for New Orleans, the New American City*. Report to the Urban Planning Committee of the Bring New Orleans Back Commission. January 11. Available at www.columbia.edu/itc/journalism/cases/katrina/city_of_new_orleans_bnobc.html.

———. 2008. *The Fork: The Floyd's Fork Greenway Master Plan*. Report to 21st Century Parks.

———. 2010. *GreenPlan Philadelphia: Our Guide to Achieving Vibrant and Sustainable Urban Places*. Report to the City of Philadelphia.

Williams Rogers, John. 1951. *Lusty Texans of Dallas*. New York: E.P. Dutton and Company, Inc.

Wines, James. 2000. *Green Architecture*. Los Angeles: Taschen Books.

Woodard, Colin. 2011. *American Nations: A History of the Eleven Rival Regional Cultures of North America*. New York: Viking Press.

Yanga, Bo and Ming-Han LI. 2012. "Ecological Engineering in a New Town Development: Drainage Design in The Woodlands, Texas." 2012. Elsevier, B.V.

Photo Credits

Anne Whiston Spirn: Fig. 4.2, 4.3

Brad Goldberg: Fig. 8.2, 8.6

Bunster-Ossa: Figs. 1.1, 2.1, 3.4, 3.5, 4.1, 4.8–4.10, 6.4, 7.2, 8.3, 9.1, 9.3, 9.4

Buster Simpson: Fig. 7.6

Daniel Capozzi: Fig. 4.7

David Witham: Figs. 5.9–5.11

Georg Schrom: Fig. 9.8

http://www.flickr.com/photos/dalbera/: 9.6

http://www.flickr.com/photos/msako23/: Fig. 5.7

James Wines: Fig. 9.7

Kevin Osborn: Fig. 9.5

Lorna Jordan: Fig. 7.7

Nando Micale: Figs. 5.1–5.6

Plains and Prairie Potholes Landscape Conservation; http://www.flickr.com/photos/plainsandprairielcc/: Figs. 5.8, 5.9

Stuart Collection, University of California, San Diego: Fig. 7.1

Tim McDonald: Figs. 10.2, 10.3

WRT: Figs. 1.2–1.3, 3.1–3.3, 4.4, 6.1–6.3, 7.3–7.5, 8.1, 8.4, 8.5, 8.7, 9.2, 10.1

Index

AARP, Complete Streets program, 63
abstract-concrete continuum, 68, 78. *See also*
   mind-body continuum
abstraction. *See also* ethics of design;
   public art
   Dallas revitalization, 132–36, 142–43
   design value of, 83
   Fargo 365 project, 80–81
   Georgetown Waterfront Park project,
      104–6
   Louisville park design, 151
   McHarg pedagogy, 84, 86
   Miami park design, 97–98
   Santa Monica Beach Improvement Group
      project, 119–21
   Broward County art plan, 126
Acadia National Park, 29
active mobility, 63–65, 155, 158, 176
African Americans, urban sprawl affecting, 10–12,
   51
airports, protected wildlife near, 20
Alameda County. *See* South Livermore (Calif.)
Alexander, Christopher, 135
allegory. *See* abstraction; ethics of design; public art

Almy, Dean, 88
American Disability Act, 157
American Heart Association, Complete Streets
   program, 63
American Jobs Act, 177
American Nations (Woodard), 91
American Planning Association, Complete
   Streets program, 63
American Samoa, wildlife refuge in, 21
American Society of Landscape Architects,
   Complete Streets program, 63
Anacostia community (Washington, D.C.). *See
   also* Washington (D.C.)
   green architecture in, 161
   river revitalization project, 11, 12, 98–99
Anacostia River. *See* Washington (D.C.)
Anacostia Waterfront Initiative, 98–99
Andersen, Martin A., 33
Appalachia, ecology of, 45–46
Appleyard, Donald, *Temporary Paradise?: A Look
   at the Special Landscape of the San Diego Region*,
   150
Arches National Park, 29

193

Arizona
   national parks in, 23
   protected land in, 38
   urban density in, 41, 42
art. *See* public art; unity of the arts
artists, designers' opinion of, 110–12
Atlanta (Ga.)
   density of, 41
   project funding in, 174
   smart growth in, 38, 172
   urban densification in, 42
automobiles. *See also* highways; streets
   emissions from, 56–57
   result of dependence on, 32–33
   traffic calming measures, 58–59

Baca, Judith, *The Great Wall of Los Angeles*, 122
*Badland*s, 25
Bakersfield (Calif.), wilderness near, 27–28
Baltimore (Md.). *See The Plan for the Valleys*
Baroque art and architecture, 68–79, 86
Bartholomew, Harlan, 131–32
Basin and Range Province, 27–28
Beach Improvement Group (BIG) Project, 117–21
Beckman, John, 152
Beckoning Cistern (Simpson), 124, 125
Berger, Alan, 179
Bernini, Lorenzo
   Fontana dei Qiuattro Fiumi, 69–71, 75, 77
   Sant'Andrea al Quirinale, 71, 73, 75, 77
bicycles
   Dallas planned trails for, 145
   greenways for, 158
   Philadelphia planning for, 63–65
   public health and, 33

   regional trail for, 103
Bierstadt, Albert, *Looking Down Yosemite Valley*, 23
BIG (Beach Improvement Group Project), 117–21
Big Bend National Park, 29
"Bio Boulevard and Water Molecule" (Simpson), 124
biocentrism, 14, 29
Bismarck (N. Dak.), regional identity of, 91–92
body, mind connection to. *See* mind-body continuum
Borromini, Francesco
   Palazzo Barberini staircase, 75, 76, 77
   San Carlo Alle Quatro Fontane, 71, 74, 75, 77
   Sant'Agnese in Agone, 69, 75, 77
   Sant'Yvo della Sapienza, 71, 75, 77
Boston (Mass.)
   Big Dig project in, 87–88, 93
   density of, 41
   public meetings in, 93
   regional identity of, 92
   return on green development in, 173
Bradford, William, 22
Bravura, 151–52
Bressi, Todd, 128
Brookner, Jackie, 122
Broward County (Fla.), Environmental Art Master Plan, 124, 126–28
Bryan, John Neely, 142–43
Buffalo Commons initiative, 15
building, as garden, 148, 160–66
Bureau of Public Roads, 27
Burlington County (N.J.), public meetings in, 93
Burns, Carol, 87, 88
Bush, George H.W., 132

INDEX

Cabeza de Vaca, Álvar Nuñez, 22
Calatrava, Santiago, 136, 144
California. *See also specific cities*
   highway system, 27, 168
   housing trends, 168–69
   protected land in, 38
   University of, at San Diego, 113–15
   wilderness in, 27–28
Camden (N.J.), public meetings in, 93
carbon emissions. *See also* microclimate mitigation
   in GreenPlan Philadelphia, 55–59
   highway planning and, 10
   "net zero," 170
   in Trinity River Corridor Project, 173
   vegetative wall managing, 47
   walkability and, 35
Carrolton (Tex.), Dallas connection to, 131
Carson, Rachel, 19, 29–30
   Silent Spring, 16
Cassirer, Ernst, 84
Catholic Church, Baroque influence of, 78, 86
Caves Valley (Md.). *See The Plan for the Valleys*
Centers for Disease Control and Prevention, 33
Central Pacific rail company, 25
CFA (Commission of Fine Arts), 101, 104, 105
Chan Krieger & Associates, 137
Chaney, Rufus, 122
Charlotte (N.C.), public art in, 110
Chicago (Ill.), density of, 41
Chiesa de Sant'Ignazio (Rome), 71, 72, 75, 77
Chin, Mel, *Revival Field*, 122–23
Christ of Saint John of the Cross (Dali), 119
Christo and Jeanne-Claude
   The Gates, 111
   Surrounded Islands, 111

CH2M Hill, 137
cities. *See also specific*
   communities linked to, 158
   concrete and abstract in designing, 85
   effect of sprawl on, 10–13
   as landscapes, 148, 149–54 (*see also* Climax City)
   public involvement in planning (*see* community engagement process)
   as unfit, 5, 13
   unity of the arts in designing, 88–89
Civil Rights Act, 11, 12
Clean Water Act, 174
climate, preserving. *See* microclimate mitigation
Climax (Mich.), steady state urban condition of, 147
Climax City, 147–66
   building as garden, 148, 160–66
   city as landscape, 148, 149–54 (*see also* abstraction)
   community as park, 148, 155–60
   densification in, 168–72, 177
   requirements for, 147–48, 176–77
   unity of the arts for, 179
Clovis (N.Mex.), Combined Statistical Area, 32
Combined Statistical Area (CSA), 32
Commission of Fine Arts (CFA), 101, 104, 105
community. *See also specific locations*
   future of, 168
   as park, 148, 155–60
   smart, 169
community engagement process. *See also* localism
   examples of, 95–99, 153–54
   four-step plan for, 95, 102–7
   perspectives on, 93–94
community identity. *See* localism

195

community issues, as medium for art, 122
commuting. *See also* automobiles; highways; pedestrians; public transportation
interstate highways changing, 9–10, 11
Philadelphia plan for, 63–65
without automobiles, 33, 35
Complete Streets program, 63, 155–58
compossibility, 83–86
Comprehensive Everglades Restoration Plan, 15. *See also* Everglades
conceptual virtue, 83
concrete-abstract continuum, 68, 78,85. *See also* mind-body continuum
Conestoga-Rovers Associates, 62
Congress for the New Urbanism, 34–35
Connecticut, Combined Statistical Area, 32
Coral Gables (Fla.), density of, 40–41
Corner, James, 88, 97
Crater Lake National Park, 29
CSA (Combined Statistical Area), 32
culture, in city as landscape, 149. *See also* abstraction
Cumberland (Md.), biking trail in, 103
Curtis, William, 160

da Cortona, Pietro, *Trionfo della Divina Provvidenza*, 75, 77
Dali, Salvador, *Christ of Saint John of the Cross*, 119
Dallas (Tex.). *See also* Trinity River Corridor Project (TRCP)
Balanced Vision Plan, 137
city as landscape in, 149
Dallas Visions for Community, 130, 136, 144
density of, 41, 129
highway system, 26, 27, 137–39

intercommunity greenway in, 158
railroad in, 26, 129
smart growth in, 38
urban densification in, 42, 129
Damon, Betsy, 122
The Day Before America (McLeish), 21–22
De Maria, Walter, *Lighting Field*, 111
de Souza Briggs, Xavier, 99
Democracy and Problem Solving, 94
defense, national, and U.S. highway development, 176, 177
Deleuze, Gilles, *The Fold: Leibniz and the Baroque*, 68, 78, 83, 86
Democracy and Problem Solving (de Souza Briggs), 94
densification, 34–43
for climax urban state, 168–72, 177
future of, 42–43
localism with (*see* localism)
Miami area, 39–41
population growth impacting, xi
public health and, 32–34, 49
smart growth principles, 34–38
U.S. statistics on, 31
design
ethics of (*see* ethics of design)
for fitness between man and nature, 37–38 (*see also* fitness)
open canvas approach to, 85
public involvement in, 97 (*see also* community engagement process)
design competitions, 80, 85–86
Design with Nature (McHarg), viii, xiii–xiv, xv, xix, 1–17, 167, 179
designers, opinions on artists, 110–12
Detroit (Mich.), racial tension in, 11

INDEX

disciplinary unity. *See* unity of the arts
Donner family, 28
Douglas, Marjory Stoneman, 1
Duchamp, Marcel, 109
Durkin, Martin P., 176

Earth Day, 11–12
Eco-Art, 122–28
ecological climax. *See* Climax City
ecological health, 4–5
ecological placemaking, 99–101, 168. *See also* localism
ecological planning method, 167
ecological service, xiii, 49–50, 57. *See also* working nature
ecological sustainability, 89–90
Ecological Urbanism (Mostafavi), 89
ecological utility, site specific, 121–28
ecology
    defined, 45, 67–68, 84
    as design ethic (*see* ethics of design)
    as promised land, 4–5
economic development, 143–45, 167, 172–74
education, interdisciplinary, 87
Eisenhower, Dwight D., 26, 49, 175–76
"Emo" urbanism, 89
The End of Nature (McKibben), 16
endangered wildlife, 21
Engman, Robert, 65
EPA, 56, 62, 173–74. *See also* NEPA
Erie National Wildlife Refuge, 23
Essex County (N.J.), public meetings in, 93
ethics of design, 67–90
    body-mind connection in, 68–79
    compossibility and, 83–86
    in ecological place making, 99–101

ecological relationships in, 67–68
    metaphor, allegory, and fantasy in, 79–83 (*see also* abstraction; public art)
    unity of the arts and, 86–90
Everglades, ecology of, 1–3, 15, 16
Everglades National Park, 29
evolution, fitness in, 4

fantasy. *See* abstraction; ethics of design; public art
Fargo (N. Dak.)
    city as landscape in, 149
    Fargo 365 design, 79–83, 85–86, 165–66
    regional identity of, 91, 92
Federal-Aid Highway Act, 26
Fingerspan (Pinto), 115–17
fire, in altering nature, 21–22
fitness
    concept of, 3–4
    as design goal, 4–5, 9
    success stories, 37–38
Florida. *See also* Miami (Fla.)
    art master plan in, 124, 126–28
    city as landscape in, 149–50
    density in, 39–41
    Everglades ecology, 1–3, 15, 16
    protected land in, 38
The Fold: Leibniz and the Baroque (Deleuze), 68, 78, 83, 86
folklore. *See* abstraction; ethics of design; public art
folly. *See* abstraction; ethics of design; public art
Fontana dei Qiuattro Fiumi (Bernini), 69–71, 75, 77
For Everyone a Garden (Safdie), 167
Ford, Bishop, 27

Ford, John, *The Grapes of Wrath*, 24
Fort Lauderdale (Fla.), density of, 41
Fort Worth (Tex.)
  density of, 41, 42
  highway system, 26
  foundational landscape, 150, 152, 158. *See also* Climax City
four-step public engagement process, 95, 102–7
Franks, Lawrence D., 33
Franks, Meryl, 158
Friedman, Daniel S., 96
Friedman, Thomas L., 167–68
Frisco (Tex.), Dallas connection to, 131
Full Circle (Lacy), 122
funding, 174–79

Gackle (N. Dak.), regional identity of, 91, 92
Galveston (Tex.), history of, 22
gardens. *See also* rain gardens
  buildings as, 148, 160–66
  demonstration, 52–53
  in Georgetown Waterfront Park project, 104
The Gates (Christo and Jeanne-Claude), 111
Georgetown neighborhood (Washington, D.C.), waterfront park in, 101–7, 177. *See also* Washington (D.C.)
Georgia, protected land in, 38. *See also* Atlanta (Ga.)
Ginsburg, Alan, 11
Glasgow (Scotland), urban growth in, 5
Glick, Gene, 158, 159
Glick, Marilyn, 158, 159
Goldberg, Brad, 132–33
good life, defining, vii–viii
government, and climax urban state, 174–79

The Granite Garden: Urban Nature and Human Design (Spirn), 50–53
The Grapes of Wrath (Ford), 24
Great Smoky Mountains National Park, 29
The Great Wall of Los Angeles (Baca), 122
Green Architecture (Wines), 163
green equity, 174
green infrastructure
  in densification, xii, 43
  Fargo 365 project, 80–81
  in Georgetown park project, 106
  history of Philadelphia's, 49–53 (*see also* GreenPlan Philadelphia)
  Landscape Urbanism approach to, 87–88
  New Orleans recovery and, 154
  return on investment, 172–74
  system supporting, 45–46
  urban-scale application of, 46–49
Green Spring Valley (Md.). *See The Plan for the Valleys*
Green Streets program, 155–58
GreenPlan Philadelphia, 53–65. *See also* Philadelphia (Pa.)
  active mobility, 63–65
  background, 53–55
  benefits analysis, 63–65, 67
  carbon sequestration, 55–59
  storm water management, 59–61
  watershed restoration, 61–63
Greensboro (Vt.), Combined Statistical Area, 31, 32
greenways. *See also* parks
  intercommunity, 158–60
  Louisville success story, 175
  as Philadelphia objective, 63, 65 (*see also* Philadelphia (Pa.))

INDEX

role of, in Dallas, 145

Halff Associates, 136
Halprin, Lawrence, 37
Hamilton, Alexander, vii–viii
Hammond, Stephen, 3
Hara, Mami, 53
Harbison, Robert, *Reflections on the Baroque*, 78
Hargreaves Associates, 137
Harris, Sam, 116
health. *See* public health
Heidegger, Martin, 45
Heinz, John, 19–21, 30
The High Frontier: Human Colonies in Space (O'Neil), 13
Highland Park (N.J.), street design in, 157–58
Highway Trust Fund (HTF), 176, 177
highways. *See also* automobiles
    Dallas system of, 131–32
    development history, 24–25, 26–27
    environmentally-acceptable, 9–10
    funding for U.S., 175–79
    inner cities affected by, 10–13
    Santa Monica coastal, 120–21
Homo urbis, 30, 46
Host Analog (Simpson), 123–24
The House We Live In, 1–2
Houston (Tex.)
    public transportation in, xiv–xv
    urban densification in, 42
HTF (Highway Trust Fund), 176, 177
H_2O *and the Waters of Forgetfulness* (Illich), 135–36
Hudson County (N.J.), public meetings in, 93
Hudson River School, 23
Humphry, George M., 175
Hunderwasser, Friedensreich, 164–65

Illich, Ivan, 142
    H_2O *and the Waters of Forgetfulness*, 135–36
Illinois
    highway system, 27
    urban density in, 41
Indianapolis (Ind.), intercommunity greenway in, 158–60
infrastructure. *See also* green infrastructure
    effects of sprawling, 177–78
    rising in wilderness, 24–27, 28–30
    inner city, sprawl affecting, 10–13. *See also* cities
Institute of Transportation Engineers, Complete Streets program, 63
Interstate Highway System, 26–27, 175–76. *See also* highways
investment, return on, 172–74
Iowa, Prairie pothole region, 82
Iran, environmental park in, xv, 30
Irwin, Robert, *Two Running Violet V Forms*, 113, 114
Ishii, Anna, 80

Jacobs, Jane, 135
James Corner Field Operations, 97
Jeanne-Claude and Christo
    The Gates, 111
    Surrounded Islands, 111
Jefferson, Thomas, vii–viii, 24
Johanson, Patricia, 122
Johnson, Lady Bird, 27
Johnson, Lyndon, 27, 132, 138–39
Johnson Atoll, wildlife refuge in, 21
Jones, Dan, 152, 175
Jones, David, 175
Jordan, Lorna, 122, 154
    Theater of Regeneration, 124, 126–28

Reconsidering Ian McHarg

Kahn, Louis, 51
Kahn, Ned, *Wind Veil*, 110
Kairys, Hannah Mattheus, 80
Kelley, Jeff, 110, 112
Kennedy, John F., 25, 26
Kentucky. *See* Louisville (Ky.)
Kessler, George, 130–31, 132
Kilbourne Group, 80
Kiley, Dan, 135
King, Martin Luther, 11
Kirk, Ron, 136

Lacy, Suzanne, *Full Circle*, 122
land art movement, 111, 116
landscape, city as, 148, 149–54. *See also* Climax City
Landscape Urbanism, 87–88
Lansdale (Pa.), watershed restoration, 62
Laporte (Pa.), ecology of, 46
layer cake method
  in *The Plan for the Valleys*, 7–10
  composition process versus, 100
  concept overview, 5–6
  density in, 34
  ecology and, 84
  public art link to, 126
Leibniz, Gottfried Wilhelm, 68, 83–86
Levittown (N.Y.), in national defense agenda, 176
life cycle mix, in healthy neighborhoods, 158
Lighting Field (De Maria), 111
Lincoln, Abraham, 23, 25
Lincoln Institute of Land Policy, 168
Lippard, Lucy R., 113
  The Lure of the Local: Senses of Place in a Multicentered Society, 91, 93, 101
living streets, 155–58

localism, 91–107
  community engagement, 94–99
  in densification, xii, 43
  in design concept choice, 83
  as "eco-democratic" circumstance, 93–94
  ecological placemaking, 99–101, 168
  in public art, 112–13
  regional identities, 91–92
  urban waterfront park project, 101–7
Lombardy (Italy), building as garden in, 160–61
Looking Down Yosemite Valley (Bierstadt), 23
Los Angeles (Calif.)
  racial tension in, 11
  smart growth in, 38
  water diversion plan, 15–16
Louisiana, post-hurricane rebuilding in, 152–54
Louisville (Ky.)
  city as landscape in, 150–52
  project funding in, 175
Lowe, Rick, *Project Row Houses*, 122
The Lure of the Local: Senses of Place in a Multicentered Society (Lippard), 91, 93, 101
Lynch, Kevin, *Temporary Paradise?: A Look at the Special Landscape of the San Diego Region*, 150

MacArthur, Douglas, 27
Mackie, Jack, 122
MacLeish, William, *The Day Before America*, 21–22
Malaysia, green architecture in, 161
Malick, Terrence, 25
MAP-21 (Moving Ahead for Progress in the Twenty-First Century Act), 176
mapping, pre-computer, 6
marshes, dwindling, 19–21
Martí, José, 97

Index

Maryland, biking trail in, 103. *See also The Plan for the Valleys*
Massachusetts. *See* Boston (Mass.)
McGuire, Mark, 26–27
McHarg, Ian
    on artists, 110–11
    building on legacy of, 179
    death of, xix
    deification of nature, 13
    Design with Nature, viii, xiii–xiv, xv, xix, 1–17, 167, 179
    at Earth Day celebration, 11
    television appearance by, 1–2
    in WMRT formation, 86 (*see also* WMRT)
McKay, Douglas, 175
McKibben, Bill, *The End of Nature*, 16
McKinney (Tex.), Dallas connection to, 131
Meehan, Doug, 80
mental health, as green initiative goal, 130. *See also* abstraction; ethics of design
Mercer County (N.J.), public meetings in, 93
metaphor. *See* abstraction; ethics of design; public art
metropolitan areas, 32. *See also* cities
Metz Engineers, 62
Miami (Fla.)
    density of, 39–41
    public art in, 109
    riverfront park project in, 97–98
Miccosukee tribe, 1
Michigan
    racial tension in, 11
    steady state urban condition in, 147
microclimate mitigation, 82, 140. *See also* carbon emissions
Micropolitan areas, 32

Middlesex County (N.J.), public meetings in, 93
Midway Atoll, wildlife refuge in, 21
Mill Creek Housing Project, 51. *See also* Philadelphia (Pa.)
Miller, Laura, 137
mind-body continuum, 68, 69, 71, 78, 79. *See also* abstraction; Baroque art and architecture; ethics of design
Minnesota
    prairie potholes in, 82
    public art in, 122–23
Missouri, highway system in, 26–27
mitigation, microclimate, 82, 140. *See also* carbon emissions
mixed use development, 39–41, 42, 170–72
Mobil Oil, 133
Mobius strip, 45, 163
mode-sharing percentage, 63–64
monads, universe composed of, 83–84
Moore, Charles, 37, 135
Moore, Hiram, 147
Mostafavi, Mohsen, *Ecological Urbanism*, 89
Mount Vernon (Va.), biking trail in, 103
Moving Ahead for Progress in the Twenty-First Century Act (MAP-21), 176
Mulholland, William, 15–16
Muskie, Edmund, 11
mythology, as design inspiration, 132–35. *See also* abstraction

Nader, Ralph, 11
Nash, Roderick, *Wilderness and the American Mind*, 22
National Capital Planning Commission, 102
National Environmental Policy Act, 10

National Environmental Protection Act (NEPA), 93–94. *See also* EPA
national grid, 24–25. *See also* highways
National Interstate and Defense Act, 26
National Park Service, 101, 103, 104, 149
national parks, 23, 29, 30, 101
National Tree Benefit Calculator, 173
National Trust for Public Land, 172–73
National wildlife refuges, 19–21, 30
Native Americans, and nature, 1, 21–22
nature. *See also* wilderness
  in city as landscape, 149
  consideration of remaining, 16–17
  humankind's relationship with, 13–17
Nauman, Bruce, *Vices and Virtues*, 113, 115
Nebraska, railroad in, 25
Nelson, Gaylord, xi
NEPA (National Environmental Protection Act), 93–94. *See also* EPA
"net-zero" development, 170–72
Neukrug, Howard, 53
New Genre public art, 122
New Jersey
  Combined Statistical Area, 32
  highway system, 10
  public meetings in, 93
  racial tension in, 11
  street design in, 157–58
  suburban planning in, 9
New Mexico
  Combined Statistical Area, 32
  missile range in, 23
New Orleans (La.), post-hurricane rebuilding in, 152–54
new urbanism, 34–35, 168

New York City (N.Y.)
  Central Park role in, 155
  public art in, 111
New York State
  Combined Statistical Area, 32
  and national defense, 176
Newark (N.J.), racial tension in, 11
Noguchi, Isamu, 109
North Dakota
  city as landscape in, 149
  Fargo 365 design, 79–83, 85–86, 165–66
  highway system, 25
  Prairie pothole region, 82
  regional identities in, 91–92
nuclear waste, 28
Nutter, Michael, 53

Oakland (Calif.), viaduct collapse in, 94–95
Obama, Barack, 28, 176, 177
obesity, and city design, 33, 35
Office of Management and Budget (OMB), 32
Oldenburg, Claes, 65
Paint Torch, 109
Olin Partnership, 61
olive groves, as open space, 37
Olmsted, Frederick Law, xii–xiii, 152–54
Olmsted, Frederick Law, Jr., 12
Omaha (Neb.), railroad in, 25
OMB (Office of Management and Budget), 32
O'Neil, Gerard K., *The High Frontier: Human Colonies in Space*, 13
Onion Flats, 171
open space. *See also* greenways; parks; *specific locations*
  in community as park concept, 155
  conservation of, 35, 36, 38
  working nature as, 37, 49

Index

Oregon. *See* Portland (Ore.)
Orlando (Fla.), city as landscape in, 149–50
Osburn, Kevin, 159

Pacific Islands, wildlife refuge in, 21
Paint Torch (Oldenburg), 109
Palazzo Barberini (Rome), 75, 76, 77
Pardisan environmental park (Iran), xv, 30
parks. *See also* greenways; *specific cities*
    community as, 148, 155–60
    national, 23, 29, 30, 101
    National Park Service, 101, 103, 104, 149
    return on investment from, 172–73
participatory ecology. *See* localism
Payne, Bryan, 158
pedestrians
    Philadelphia plan for, 63–65
    public health and, 33, 35
    street design friendly to, 155–58 (*see also* streets)
Pegasus, as design inspiration, 132–33, 134
Penn, William, 65
Pennsylvania. *See also* Philadelphia (Pa.)
    Combined Statistical Area, 31, 32
    ecological service in, 46
    marshes in, 20
    watershed restoration in, 62
Peter Walker & Partners, 149
Philadelphia (Pa.)
    building as garden in, 172
    Combined Statistical Area, 31, 32
    community as park in, 155, 156, 172
    ecological perspective, 34
    green infrastructure history, 49–53 (*see also* GreenPlan Philadelphia)
    greened-acres program in, 173–74

"net-zero" development in, 170–72
public art in, 109, 115–17, 171
public meetings in, 93
water conservation plan, 60–61
wildlife refuges, 19–21, 23, 30
Wissahickon Valley Park, 115–17
Phoenix (Ariz.)
    density of, 41
    smart growth in, 38
Pinto, Jody
    Fingerspan, 115–17
    Georgetown project, 101, 104, 105, 106, 107
    Santa Monica project, 117–20
placemaking, ecological, 99–101, 168. *See also* localism
The Plan for the Valleys
    area map, 6
    multi-disciplinarity of, 86–87
    open space conservation, 38
    planning process, 7–10
Plano (Tex.), Dallas connection to, 131
PLSS (Public Land Survey System), 24–25
Poetic Utility (Simpson), 124
population, urban percentage of, 31–32
Portalis (N.Mex.), Combined Statistical Area, 32
Portland (Ore.)
    active mobility in, 64
    public art in, 123–24, 125
Potomac River. *See* Washington (D.C.)
prairie pothole region, 79–83
Pratt, James, 130
private funding, 175
productivity, national, and U.S. highways, 176
program, in ecological place making, 99–101
Project Row Houses (Lowe), 122
promised land, ecology as, 4–5

203

Reconsidering Ian McHarg

Promontory Summit (Utah), train tracks in, 25
public art, 109–28. *See also* unity of the arts
   architects' view of, 110–12
   Baroque, in Rome, 68–78
   in Dallas revitalization, 132–33, 134, 141–43, 145
   in densification, xii, 43
   in design concept choice, 83
   in Georgetown park project, 102–7
   GreenPlan Philadelphia and, 65
   on Indianapolis greenway, 159–60
   metaphor, allegory, folly, and fantasy in, 109 (*see also* abstraction; ethics of design)
   in places of refuge, 115–21
   site specificity, 112–15, 121–28
   on streets designed for living, 155–58
   value of, 128
   wilderness depicted in, 23
public health
   active mobility and, 63–65, 155, 158, 176
   stress relief, 130
   urban living link to, 32–34, 49
public involvement. *See* community engagement process; localism
Public Land Survey System (PLSS), 24–25
public meetings, role of, 93. *See also* community engagement process; localism
public transportation
   availability of, xiv, 35
   cost of, 176, 178
   Dallas system of, 129
   Puerto Rico, wildlife refuge in, 21

race, urban sprawl and, 10–13, 51
Rahenkamp Sacks Wells Associates, 9

railroads, development history of, 25–26. *See also* public transportation
rain gardens. *See also* gardens; stormwater management
   Georgetown, 106
   Highland Park (N.J.), 157–58
   Indianapolis, 159
   Philadelphia, 52, 53, 170
   Portland (Ore.), 124, 125
Reflections on the Baroque (Harbison), 78
religion, in stewardship of nature, 13, 14
return on investment, in green infrastructure, 172–74
Revival Field (Chin), 122–23
Richmond (Va.), building as garden in, 163
"River of Grass," 1, 2
Riverside (Calif.), density of, 41
roads, development history, 24–25. *See also* highways
Roberts, William, xv, 8, 86
Rogers, John William, 142–43
Rogers Marvel Architects, 149
Rome (Italy), Baroque architecture in, 68–78, 86
Ross, Mindy Taylor, 160
Ruby Hill community (Calif.), 36
rural residential areas, 31–32

Sacramento (Calif.), railroad in, 25
Safdie, Moshe, *For Everyone a Garden*, 167
safe havens, access to, 30
Salem County (NJ), wildlife refuge in, 23
Salt Lake City (Utah), wilderness near, 27–28
San Bernardino (Calif.), density of, 41
San Carlo Alle Quatro Fontane (Rome), 71, 74, 75, 77

INDEX

San Diego (Calif.)
   city as landscape in, 150
   multifamily buildings and, 169
   return on green development in, 173
San Francisco (Calif.)
   city as landscape in, 149–50
   green architecture in, 161
   urban living trend, 169
Sandel, Michael, ix
Santa Monica (Calif.)
   beach improvement project, 117–21
   Palisades Park project in, 98
Sant'Agnese in Agone (Rome), 69, 75, 77
Sant'Andrea al Quirinale (Rome), 71, 73, 75, 77
Sant'Yvo della Sapienza (Rome), 71, 75, 77
Saunders, Russ, 119
Scarpa, Carlo, 160, 161
Schmid, Thomas L., 33
Schulz, Christian Norberg, 135
Schuylkill River. *See* Philadelphia (Pa.)
sculpture. *See* public art
Sea Ranch community (Calif.), fitness of, 37
Seattle (Wash.)
   regional identity of, 92
   viaduct project in, 92, 94–97
Seville (Spain), World Expo architecture in, 136
sewers. *See also* stormwater management
   Philadelphia treatment of, 52–53, 59, 61, 173
   pollution from, 12
Silent Spring (Carson), 16
Simpson, Buster, 122
   Beckoning Cistern, 124, 125
   "Bio Boulevard and Water Molecule," 124
   Host Analog, 123–24
   Poetic Utility, 124

site
   in ecological placemaking, 99–101
   of public art, 112–15
   specific ecological utility of, 121–28
   specific utility of, 115–21
smart community, 169
smart growth, planning principles, 35–38
Sonfist, Alan, "Time Landscape," 111
South Dakota
   Badlands movie, 25
   Prairie pothole region, 82
South Livermore (Calif.), design of, 3, 35–38
Spirn, Anne Whiston, 13
   The Granite Garden: Urban Nature and Human Design, 50–53
sprawl
   affecting inner city, 10–13, 51
   economics and environmental impacts of, 167
   mixed use in preventing, 42
   water project leading to, 16
St. Louis (Mo.), highway system, 26–27
St. Paul (Minn.), public art in, 122–23
Stanford, Leland, 25
stewardship, of nature by humankind, 13–17
stormwater management. *See also* rain gardens
   as Green Streets focus, 155
   "green" versus "gray" approach to, 174
   GreenPlan Philadelphia, 59–61
   in New Orleans recovery plan, 152, 153
   public art in, 124
   Texas plan for, 129, 138, 144 (*see also* Trinity River Corridor Project (TRCP))
Street, John, 53
streets
   bicycle lanes in, 64–65
   Complete Streets program, 63, 155–58

Reconsidering Ian McHarg

Green Streets program, 59, 155–58
tree planting along, 58–59
stress, green space relieving, 130. *See also* abstraction; ethics of design
suburban living, side effects of, 32–33
suburbs, evolution of, viii–ix
Supawna Meadows National Wildlife Refuge, 23
Surrounded Islands (Christo and Jeanne-Claude), 111
sustainability. *See* green infrastructure

Taiwan, environmental park in, 30
Taroko National Park (Taiwan), 30
Tempe (Ariz.), urban densification in, 42
Temporary Paradise?: A Look at the Special Landscape of the San Diego Region (Appleyard and Lynch), 150
Ten Architectos, 141
Texas. *See also* Dallas (Tex.)
  density in, 41, 42
  early exploration of, 22
  highway system, 26
  protected land in, 38
  public transportation in, xiv–xv
  University of, at Arlington, 129
  urban sprawl in, xiv–xv
  Vision North Texas project, 129 (*see also* Trinity River Corridor Project (TRCP))
Theater of Regeneration (Jordan), 124, 126–28
Thomas, Gail, 130, 132, 135, 137
threatened wildlife, 21
"Time Landscape" (Sonfist), 111
Tinicum Marshlands, 19, 20–21, 23
TKF Foundation, 106
Todd, Tom, xv, 86
Touch Sanitation (Ukeles), 122

traffic calming, 58–59. *See also* automobiles; highways; streets
training, interdisciplinary, 87
trains. *See* public transportation
Transcontinental Railroad, 25
transfer of development rights, 7–8, 36
transit. *See* public transportation
TRCP. *See* Trinity River Corridor Project (TRCP)
trees
  carbon sequestration by, 173 (*see also* carbon emissions)
  in Dallas area river development, 140
  in Everglades, 1, 2
  Philadelphia planting program, 55–59
Tregoning, Harriet, vii–x
Trinity River Corridor Project (TRCP), 129–45. *See also* Dallas (Tex.)
  abstraction in, 132–36
  carbon sequestration, 173
  ecological climax in, 148
  economic development, 143–45, 172
  environmental land use, 139–40
  history of, 129–30
  parkway design, 137–39
  planning background, 130–32
  project funding, 174–75
  public art in, 132–33, 134, 141–43, 145
  recreation in, 140–41
  steps in, 136–41
Trionfo della Divina Provvidenza (Rome), 75, 77
Truman, Harry, 29
Trust for Public Land, 38
Tuan, Yi Fu, 135
Turnbull, William, 37–38
Twain, Mark, 26–27

Index

Two Running Violet V Forms (Irwin), 113, 114

Ukeles, Mierle Laderman, Touch Sanitation, 122
Unger, Paula, 119
Union County (N.J.), public meetings in, 93
Union Pacific rail company, 25, 26
unity of the arts. *See also* public art
    Baroque style and, 78, 121
    in building as garden concept, 161
    concept overview, 86–90
    for future climax urban state, 179
    success story, 121
Universal Paragon Corporation, 169
University of California at San Diego, 113–15
University of Texas-Arlington, 129
urban ecology
    focus on, xii, 13
    goal of, 68
    localism in, 97, 112, 129
    McHarg successor on, 50 (*see also* Spirn, Anne Whiston)
urban housing. *See also specific locations*
    future of, 168–72, 177
    multifamily, 41, 164, 178
    public housing projects, 51
    U.S. percentage of, 31–32
Urban Land Institute, 129, 170
urban living. *See also specific locations*
    human health and, 32–33
    need for green spaces in, 130 (*see also* abstraction; ethics of design)
    U.S. statistics, 31–32
urban nature, xii, 17, 43, 110, 128
U.S. Army Corps of Engineers, 137, 175
U.S. Department of Agriculture, 31
U.S. Fish and Wildlife Service (FWS), 21, 23

U.S. Geological Survey, 27

Venice (Italy), building as garden in, 161, 162
Venturi, Robert, 135
Vermont, Combined Statistical Area in, 31, 32
vertical development, 39–41
Vices and Virtues (Nauman), 113, 115
Vienna (Italy), building as garden in, 162
vineyards, as open space, 37
Virgin Islands, wildlife refuge in, 21
Virginia
    biking trail in, 103
    building as garden in, 163

Waldheim, Charles, 88
walkability, 35, 158. *See also* pedestrians
Wallace, David, xv, 3, 7, 8, 39, 86, 102
Washington (D.C.)
    Anacostia neighborhood in, 11, 12, 98–99, 161
    Constitution Garden design in, 149
    Georgetown neighborhood in, 101–7, 177
    green architecture in, 161
    highway system, 11
    public meetings in, 93
    return on green development in, 173
    river revitalization project, 11, 12, 98–99
    sprawl surrounding, 177–78
    urban waterfront park in, 101–7, 177
water, urban management of. *See* stormwater management
Water Allocation Resources Act, 175
watershed restoration, 61–63. *See also* stormwater management
Weeks, Sinclair, 176
White, William H., 135
White Sands Missile Range, 23

wilderness, 19–30. *See also* nature as boundless, 22–23
  infrastructure rising in, 24–27, 28–30
  largest remaining, 27–28
  Native Americans altering, 21–22
  preservation legislation, 23
  wildlife refuges, 19–21, 23–24, 30
Wilderness and the American Mind (Nash), 22
wildlife
  in Prairie pothole region, 79, 82
  public art impact on, 111
  refuges for, 19–21, 23–24, 30
  in river development planning, 140
  threatened or endangered, 21
Wind Veil (Kahn), 110
Wines, James, 161, 163
  Green Architecture, 163
Witham, David, 80
WMRT, xv, 86–87, 102
Woodard, Colin, *American Nations*, 91
The Woodlands (Tex.), concept behind, xiv–xv
working nature. *See also* ecological service
  in GreenPlan Philadelphia, 54–55
  open space as, 37, 49
  in public art, 122–23

Worthington Valley (Md.). *See The Plan for the Valleys*
WRT
  BIG project collaboration, 118
  Dallas Trinity River project, 136, 137
  evolution of, xv
  Fargo 365 project, 80
  GreenPlan Philadelphia, 54, 61–62
  Highland Park streetscape, 158
  Louisville park project, 151
  Miami development project, 39–41
  New Orleans recovery plan, 152–54
  The Plan for the Valleys update, 8
  Seattle viaduct project, 95–96
  South Livermore plan, 3, 36–38

Yeang, Kenneth, 161
Yellowstone National Park, 23